中国生活方式
A Unique Chinese Experience

健壹营造志

The Story of
JE Mansion

康健一 ◎ 著

中信出版集团·北京

序一

马未都

按说烹饪与建筑是两件不搭界的事，可偏就有人把这两件事完美地结合在一起。在赏心悦目的建筑环境中用上一餐美味佳肴，让色香味与形光线融为一体，此时方知饭菜之色香味并非仅口舌之快，而建筑之形光线也并非一时视觉之娱。

北宋有个建筑学家叫李诫，字明仲，家学渊源，至少四代人在朝廷为官。北宋时期，社会渐渐富裕，盖房者众，李诫作为将作监开始为宫廷做事，把握建房品质与原则。久而久之，对营造之事了如指掌；绍圣四年（1097年）李诫奉旨编修《营造法式》一书，历三年而成，于元符三年（1100年）刊行，全书三十四卷，为宋时营造房屋官方之规范。此书影响后世至今，可称营建鼻祖。

清代乾隆时期有个文人叫袁枚，乾隆四年进士，为官知县，四十岁时告归在江宁（今南京）筑随园，随遇而安。袁枚文人之名不如其美食之名，《随园诗话》不如《随园食单》让人津津乐道。袁枚以其文人细腻之笔触，将乾隆时期富庶江南的烹饪技巧不露声色地记录于纸上，凡菜肴326种，蔚为大观。

李诫与袁枚间隔七百年，按说两个人也没有任何关联，唯一强拉硬扯的关系就是两人的著作都长时期地作为后世之范本，凡建筑凡烹饪都以此为据，享其中之乐，得其中之精。搞建筑的不懂《营造法式》，学烹饪的不知《随园食单》，会成为行业笑柄，愧对祖宗。

北京东西有两个健壹，东为健壹公馆，西为健壹景园。两处健壹的主人乃为一人，说其是建筑设计家，凡来用餐的宾客又尽享烹饪之绝；说其是烹饪美食家，又让饮食男女们陶醉于建筑的绚烂之中；享一事得两事之乐，北京除健壹两馆绝无他处。

我与康健一先生相识多年，欣赏他做事的态度。一个人在世上混，无非做人做事，做人需要长期厮混方可知晓，而做事则窥一斑可见全豹。去健壹东西两馆，早春晚秋盛夏隆冬各有其美，融建筑于环境之中，布美食于餐台之上，一年四季不论何时，去健壹都是一份绝佳享受，都有一份人生感悟。王阳明讲知行合一，他说"知是行之始，行是知之成"，知易行难，所以朱熹早早告诫："论先后，知为先；论轻重，行为重。"看看健壹东西两馆，方能体味知行关系。康健一先生以一人之力，在当今纷杂的干扰之中，将两馆于滚滚红尘凸现清流，实属难能可贵。

建筑自古就分南北两宗，南宗北派都源于地貌环境、人文累积及四季变幻。南派建筑于青山绿水之中多显阴柔之美；而北派，春夏秋冬四季分明，阴柔、阳刚、凄寂、残酷各有其美，让美美与共，于四季分明之时，忘春夏秋冬之变，此乃健壹营建之法，把营建之法著书为志，再融入美味佳肴，算是一份功德。

精心做事，大气为人，方有此书。健壹两馆出书记录历程，可喜可贺，匆匆几言，记心于字，记情为义，是为序。

马未都

丁酉小满

Foreword I

Ma Weidu

The Creation of JE Mansion

Cuisine and architecture are two rather unrelated fields, but someone just had to go and fuse them together perfectly. Enjoying a lovely meal surrounded by captivating design, where flavour and aesthetic beauty become one and the same, I have finally come to realize that eating is not only an experience for the palette, and architecture is not only an enjoyment for the eyes.

There was an architect in the Song dynasty named Li Jie, who came from a long line of accomplished scholars, among whom members of at least four dynasties served as royal officials. In the Northern Song dynasty, society gradually became more prosperous, thus more and more houses being built, at which point Li Jie was selected by the royal court to oversee the quality and principles of home construction. Over time, he became a master of the process. In 1097 (4th year of the rule of Emperor Shaosheng), Li Jie, upon imperial decree, compiled the book *Treatise on Architectural Methods*, a task which required three years. Upon publication in 1100, the entire book consisted of 34 volumes, and these later acted as the official standard for architecture throughout the Song dynasty. The book's influence remains strong today, and Li Jie is considered to be one of the forefathers of classical Chinese architecture.

During the Qing Qianlong Emperor's reign there was a scholar named Yuan Mei, who excelled in Qianlong's imperial examinations at the age of only 23, upon which he was selected as an official magistrate. At the age of 40 he returned to his home in Jiangning (present-day Nanjing), where he constructed Harmony Garden, after which he settled down for the later years of his life. Yuan Mei is known less for his achievements as a scholar than as a gastronome, with his book *Poems of Harmony Garden* being considerably less widely enjoyed than his *Cookbook of Harmony Garden*. Yuan Mei used his articulate writing skill to vividly record the cooking methods of southern China during the Qianlong era on paper, including the details of an impressive total of 326 dishes.

Li Jie's time was 700 years prior to that of Li Mei, and the two had virtually nothing in common other than the fact that their works were used as reference guides for many generations to follow. This goes to show that architecture and cuisine are the same in that they are to be enjoyed and refined through time. Eventually architects who were not versed in *Treatise on Architectural Methods*, or gourmets who had not studied *Cookbook of Harmony Garden* in depth were laughed at by others in the field, and were considered disgraces to the achievements of their predecessors.

In the east and west ends of Beijing there are two JE complexes, the former being JE Mansion, and the latter JE Garden. Both of these were constructed by the same person, Kang Jianyi, who could be called either an architect or a gourmet, as those who visit these venues can enjoy the very best of both worlds, regardless of which it is they originally come seeking. Such a conglomeration of exquisite architecture and fantastic cuisine can be seen nowhere else in Beijing.

I have known Mr. Kang for many years, and greatly appreciate his work ethic. One's life is fundamentally made up of one's morals and one's work ethic; the former may take much time spent with the person to understand his or her true morals and values, but to understand his or her work ethic only a small sample is needed. Both of the JE complexes are beautiful in vastly different ways in each of the four seasons; the buildings are wonderfully integrated with the environment, and the most lovely dishes are served year round, so that every visit there is a new experience, a new feast for the senses. Ming dynasty Neo-Confucian philosopher Wang Yangming said that, "Knowledge is the beginning of action, and action is the completion of knowledge." Of course, knowing is much easier than acting. As Zhu Xi reflected more than 300 years prior, "In terms of order of occurrence, knowledge comes first; in terms of importance, action takes precedent", putting in perspective the relationship between the two. Mr. Kang Jianyi, while only a single person, amidst the cacophony of contemporary society, built these two complexes up from nothing, a truly remarkable feat.

Chinese architecture is broadly divided into the northern and southern sects, with the styles of each sect originating from their respective geographical locations, environments, cultural development, and seasonal changes. Southern buildings were constructed among the verdant, rolling hills and sprawling waterways of the region, and as a result tend toward subtle, exquisite beauty; on the other hand, throughout the north, subtle, bold, desolate and even brutal environments can all be found, each with their own aesthetic qualities, further diversified by the stark transformations of the four seasons. Kang Jianyi has taken the very best of both styles of Chinese architecture, and infused his incredible structures with the greatest accomplishments of Chinese cuisine, which I believe to be a great contribution to both heritages.

This book exists because of Mr. Kang's meticulous work ethic and generous morality. I am greatly pleased to see this book published in commemoration of the completion of these two complexes, and am humbly honored to have had the opportunity to provide this foreword.

Ma Weidu
May, 2017

营造之初

序二

马旭初

十几年前，朋友说昌平有个"大宅门"，中式建筑，环境和菜品都很不错，我就去看了。一看，的确有特色。入口门头和石门墩子都是老做法，还有拴马桩。绕过影壁墙，后面是大厅，完全的四合院内景的再现，中规中矩，是大宅子的感觉。右侧大堂是传统戏园子的格局，连通二层的中庭也是老北京人听戏叫好的场子，用餐时配合曲乐演出，很是惬意。大堂外侧及过道的墙上装饰着老花板和老戏服，门扇也是收来的拆房旧货，用作隔扇，旧物新置，甚是新鲜。更为创新的是这宅门府邸里配有室内游泳池，并且将戏水的功能与传统园林里的景观营造结合在了一起，气氛是古典的，功能是现代的，这是传统手法中不曾有过的方式。我的印象里，除了民国时期清华大学的某个中式建筑和中南海有这样的做法，就只剩下昌平的这个"大宅门"了，而且这个建筑将现代的设施和传统空间结合得很好，怎不言妙呢？总结下来，这座"大宅门"里确是充满了很多老北京的皇城记忆和现代生活的体验。当年，这样精心打造的中式建筑十分罕见，我从此记住了康健一这个名字，并且有了联系。他是一个认真做事的人，如今与我这亦师亦友的关系就是从这样的一个机缘建立起来的。

多年后，健壹集团在北京已建了三处中式建筑，即宅院、公馆和景园。无论是单体的还是聚落的，建筑本身还是景观营造，我都去实地看过。巧合的是还在工地里见到正施工的我的几个徒孙，在推敲施工的细节和做法，他们认真钻研的态度我个人非常认可。对传统建筑一定要讲究，一毫一寸都不能怠慢，传承下来的造法和口诀是要不断扎实下去的。能够看到此景，对这古建筑的探索与发扬，也在回应着我多年实践的追求。总体上说，健壹集团系列的中式建筑传承了古建筑的精华，在人和景的沟通上、在古建筑的功能转型上有独特创新，把中式建筑的老规矩和西洋建筑的舒适性结合得比较好；在古风古韵上也下足了力气，激发出了中式建筑新的生命力。

健壹集团所打造的建筑基本上达到了中学为体、西学为用的标准，对明清及民国以来的建筑模式有大胆的创新和发展，也是新中国成立以后在传统建造方面的一个榜样和典范。在我看来，诚然一些细节还需细细打磨，但相较于北京的新中式建筑，这三处中式建筑表达了更为明确的对天人合一的求索。最为可贵的是，这种尊重中式建筑、重视传统文化的态度，我十分欣赏和推崇。

希望健壹集团继续努力，带动我们的城市出现更多的中式建筑精品，让我们的城市生活更加美好！

哲匠世家第十四代传人 马旭初

2014年6月30日

Foreword II

Ma Xuchu

How the Mansion Began

Ten years ago a friend told me that there was a "mansion" in Changping District, that featured traditional Chinese architecture, and both the food and environment there were great, so I went to have a look. My first impression was that the venue was indeed unique. The entrance gate and statues on either side were all made in the traditional way, and there were even horse hitching posts. Upon entering the venue, visitors were greeted by a decorative screen wall, beyond which was the spacious main hall, giving one the impression of an authentic quadrangle courtyard. The hall on the right hand was designed like a traditional Chinese opera theater, and was connected to the second floor, so that audiences could have a better view of the action. One could imagine that people could spend a very enjoyable evening there. The walls of the main hall and hallways were adorned with patterned wood carvings and old-fashioned theater costumes, while the doors were "upcycled" from old-fashioned houses, the modern building breathing new life into these rustic articles, and vice-versa. Even more unexpected was how the lower level of the mansion was fashioned into an indoor garden, and among the greenery was a swimming pool. I thought this fusion of traditional scenery and modern functionality was very innovative. This "mansion" was a combination of classicism and modernism. This venue's meticulous Chinese-style design left the name Kang Jianyi imprinted on my memory. Later, I contacted him, thus beginning our relationship as both friends and colleagues.

Years later, the JE Group built three Chinese-style structures in Beijing. I visited each of these myself, and by coincidence, several of my apprentices happened to be working at one of the sites. I asked them some questions about the project, and they answered very diligently. I think it's very important to pay attention to details in traditional architecture, and the techniques and formulas must be preserved and carried on accurately. These structures are a reflection of many years of exploration in classical architecture. JE Group shows its classical heritage, while adding elements of Western architecture and the convenience of modern living. I believe that JE has achieved this quite successfully, and has brought new energy to traditional Chinese architecture.

JE's works are Chinese in foundation and Western in execution, and the team have made impressive innovations in architectural styles from the Qing dynasty and Republican period. The structures serve as models and paradigms of contemporary Chinese architecture. In my opinion, a few details require further pondering on classical architecture, but in comparison to other contemporary Chinese-style buildings in Beijing, these three structures more closely adhere to the concept of "the coexistence of man and nature", and I greatly appreciate their respect toward Chinese architecture and emphasis on traditional culture.

I hope that JE Group continues to work hard to create a greater number of such Chinese structures, so that Beijing and our lives there may become even greater.

Ma Xuchu
June 30, 2014

Ma Xuchu, 14th generation of architectural craftsmen

目录
Contents

4
序一　马未都

8
序二　营造之初　马旭初

6
Foreword I Ma Weidu

10
Foreword II How the Mansion Began Ma Xuchu

1
开篇

18

1
Opening

18

2
建筑作品篇

46

昌平垄上　迎祥宅院
朝阳静颐　健壹公馆
远黛亭堂　西山景园

2
Architecture Projects

46

JE Changping Manor
JE Chaoyang Mansion
JE West Mountain Garden

3
工筑文法要义篇

108

大木构架
砖瓦土石
色彩漆画
内外格局

3
Styles and Techniques

108

Wooden Framework
Brick, Tile, Clay and Stone
Color and Lacquer
Interior and Exterior Layout

4
长物赏析篇

150

山水景观
户牖之艺
家具陈设
绿植花木

4
Ornamentations

150

Scenery
Doors and Windows
Furniture
Trees and Flowers

5
传承创新意蕴篇

198

融汇古今/沟通南北/合璧中西
哲匠精神/创意空间/精工细作/文化意蕴

5
Heritage and Innovation

198

Past and Present / North and South / East and West
The Spirit of Craftsmen / The Details / Cultural Theme

"梓人也好，工匠也罢，更称不起经商；就是按照心里的念头，本着原则，努力做到，持续做好，就成了。"

筑 / 著者说

"I'm not a builder, or a craftsman, much less a businessman. I just do what I feel is right, follow my principles, work hard, and persevere, and that's enough."

Kang Jianyi, the architect/author

营造

建筑经营

有计划／有目的地造就

一个中国的品牌，以建筑为基础与依托，经营构筑起「令中国骄傲，让世界尊敬」的美誉环境、优雅氛围与质量生活

建筑

兴建土木工程

（比喻）建立某种感情

建筑物

说文解字／咬文嚼字：梳理本图册的脉络与内容

筑

动词：修建。如：构筑，建筑，修筑

文言：用杵捣土使实

北京城里，三座建筑的故事

昌平垄上：迎祥宅院
朝阳静颐：健壹公馆
远黛亭堂：西山景园

志

记住，记载，记录，记号
志向（追求／理想），志趣（兴趣／思想），
志愿（愿望）

一位筑（著）者与他的团队，及其作品，
十三年间的轨迹与演进

1

开 篇

Opening

建筑，人类的居所，凝固的文明，历史的发展，使用的衍变，艺术的载体，技法的体现。许多典故，许多传奇。

建筑，作为一种复杂的工程，涉及木、砖、瓦、石、土等许多材料、工种和技法。我们中国有很多建筑故事。从先秦时期开始，就有关于专门从事建筑设计的匠师的记载，无论是唐朝文学家柳宗元的《梓人传》，还是清朝"样式雷"那梁木入榫与皇帝行礼同步的传说，或是流传至今的成语"班门弄斧"，可以溯源到春秋时期的巧匠鲁班，这些都反映着建造、工筑与我们中国人生活的息息相关，妙趣横生。

20世纪40年代，中国建筑师、建筑史学家和教育家梁思成先生在其《图像中国建筑史》一书中，曾这样写道："中国的建筑是一种高度'有机'的结构。……如今，随着钢筋混凝土和钢架结构的出现，中国建筑正面临着一个严峻的局面。诚然，在中国古代建筑和最现代化的建筑之间有着某种基本的相似之处，但是，这两者能够结合起来吗？中国传统结构体系能够使用这些新材料并找到一种新的表现形式吗？可能性是有的。但这决不能盲目地'仿古'，而必须有所创新。"实践是检验真理的标准之一，我们欣喜和幸运地发现，在北京出现了这样的"有机"建筑作品，其设计和建造者秉承传统，孜孜以求，勇敢创新而持续耐久的探索和实践，为我们所居住的环境，筑立了传承与创新，融合与交汇的建筑作品；同时也引亮了"理想照入现实"的建筑探索实践。

健壹团队，从2003年开始至今，13年，汲取传统建筑精神，合理运用现代技术，又关照到现代生活方式，打造出具有中国韵味，有机结合了古典与现代，融合汇通了地域与文化意蕴的三座别致建筑作品（昌平宅院，健壹公馆，健壹景园），可谓是在以上所述的建筑挑战中脱颖而出的杰出代表。

健壹景园写生
Scenic painting of JE Garden

Buildings act as not only the living spaces for humankind, they also manifest civilization and history, and reflect the changes in our lives, art and technique. So many legends, so many stories.

Architecture is a complex type of engineering project, involving a wide variety of materials and techniques. The story of China's architecture begins in pre-Qin times, from which records of specialized architects have been found. Many traditional styles are still used in China today, and there are also expressions in the Chinese language referring to legendary craftsmen, thus reflecting the importance of architecture in Chinese culture and society.

In the 1940s, Chinese architect, architecture historian and educator Liang Sicheng wrote in his book *A Pictorial History of Chinese Architecture*: "Today, with the appearance of reinforced concrete and steel frame construction, Chinese architecture is facing a grave conundrum. Indeed, there are some similarities between ancient and modern Chinese architecture, but can the two be integrated? Can traditional architecture use new materials to find novel expressions? The possibility exists. However, it can by no means be merely 'mimicking antique style', it must involve innovation." Practice is a standard for proving truth, and we are fortunate to have these organic buildings in Beijing, whose designs fuse tradition with innovation, to create living spaces that are works of art in their own right.

Since 2003, Kang Jianyi and his team created three architectural works, bring traditional aesthetics and modern technology together, created structures that can be enjoyed for their beauty, while also providing the practicality of urban living. The three extraodinary venues represent the very best of modern innovation in traditional Chinese architecture.

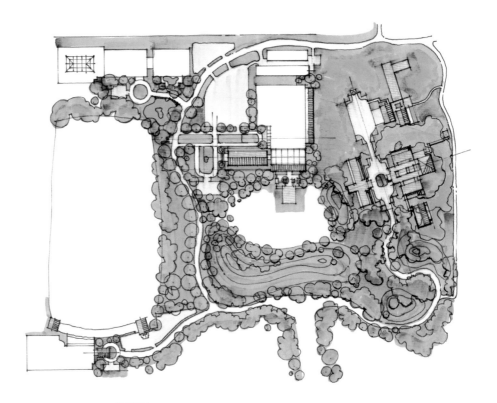

健壹景园总平面图（手绘）
Hand-painted plan of JE Garden

景园
The Garden

健壹景园礼堂立面形如牌楼
Facade of the JE Garden hall, designed to resemble an ornate archway

The Story of JE Mansion

23

健壹景园入口外
Outside of the entrance to JE Garden

健壹景园礼堂侧面
The side wall of the JE Garden hall

健壹景园书苑门,银装素裹飘芸香
Shuyuan Gate of JE Garden, coated with fresh snow

The Story of JE Mansion

玉泉山塔背景
Yuquan Mountain Pagoda can be seen in the background

公馆
The Mansion

健壹公馆门厅
JE Mansion lobby

健壹公馆南侧静谧的水池
Tranquil pond on the south side of JE Mansion

健壹公馆270度视角
270 degree views of JE Mansion

健壹公馆门廊
JE Mansion covered entrance

The Story of JE Mansion

35

宅院
The Manor

迎祥大宅门外景
Exterior of JE Manor

宅院东门
JE Manor east gate

宅院厅堂
JE Manor great hall

左图：楼上楼下，戏里戏外
The left: The audience may enjoy Beijing opera from either floor

四季
物候
The four seasons

春和景明，碧桃漫坡
The warm and clear spring, fresh blossoms and still waters

四季
物候
The four seasons

夏

夏荷涟漪，亭阁映波
Summer lotuses, the tranquil pavilions
reflected in the gentle ripples of the ponds

四季
物候
The four seasons

秋

菩提秋风，黄叶满地；雨后庭院，水气清新
The crisp autumn wind sheds the golden leaves from the trees, spreading them across the ground, and freshening the air

四季
物候
The four seasons

西山晴雪，暖阁书苑
A sunny day on snow-covered West Mountain, the library warm and cozy

指挥若定,梓人遗风
The craftsman guides the construction of the project

雕栏玉砌应犹在，古训砖雕重铺装
Antique carved railings and jade inlays find a new home in a contemporary setting

一砖一瓦凝聚巧匠心思，汗水铸就的艺术结晶
The craftsmen's ambitions are slowly realized, one brick and one tile at a time

建造：地理选址 建造初衷 定位
工筑：就地取材 他山之石
料材：原生原木 原始原处
技法：南北 古今 中外
长物：精致赏析

2

建筑作品篇

Architecture Projects

健壹和他的团队，从2003年开始，至2016年，13年间，先后在首都京城的北（昌平）、东（朝阳）、西（海淀）三个区域，建造了三座风格不同，却秉承中式建造精髓，蕴含别致韵味的建筑作品。

昌平垄上　迎祥宅院

2003年健壹品牌初创，将北京昌平的一幢旧楼，改造成富含北方特色、京城韵味的宅院，备有客房居住、餐聚雅集、观戏品茗等功能与设施。

昌平宅院突破了旧舍原有的格局限制，把整座四合院落、一座戏楼舞台和相关配套设施，巧妙而完整地含纳于一座宅府合院的建筑之中，每一厅房、连廊、院落，都因功能规划而被赋予独特的内外空间设计和搭配，以京韵为基底，收纳配饰了北方地区的传统建筑元素，营造出颇具中国北方风格的宅院文化氛围。正如其院落门联所书："忠厚传家久，诗书继世长"。

说到宅院，一般会参照中国古代居住建筑大体分为宫、府、宅三个等级。"宫"指皇族宫廷建筑，比如紫禁城、颐和园；"府"指王公贵族府邸，比如恭王府、贝勒府等；"宅"则指其余官员和百姓的宅院住所，其中上等宅院被称为"宅门"。

而老北京最典型的宅子就是四合院。四合院由大门、倒座房、垂花门、正房、厢房、后罩房等不同房屋围合而成，并且根据主人的身份和住宅的规模可以演变出各种繁简不同的组合形式。方正对称的格局，尊卑有序的空间，浑厚敦实的风格，浓荫遮蔽的院落，成为北京人传统意义上最推崇的理想生活场所。

仔细审视一下这坐落在昌平城区最繁华街道府学路上的健壹迎祥大宅门，其素朴稳健、巧妙构思的改造，将一座旧楼脱胎换骨成为充满京味京韵的端庄典雅宅院，且蕴含着传统四合院的空间意趣和文化精髓。

府学路是昌平最繁华的一条街道，附近有中国政法大学、中国石油大学、国防大学等著名学府，堪称风水宝地。健壹迎祥大宅门酒店位于街北，南面墙上雕有"大宅门"三字。正门为五开间卷棚顶硬山建筑，前立两座金色狮子，外檐一排大红灯笼之下开启六扇板门，柱上楹联"门开喜迎吉祥客，宅旺高坐有缘人"，嵌有"宅"、"门"二字。东南侧另有一间旧式广亮大门，门前立两根拴马桩，门上对联为"诗书继世，忠厚传家"——恰如其分地形容了宅院主人营造的初衷与宅门氛围。

宅院的改造设计有不少妙笔特色，选几点代表之处，一一道来。

⊙ 影壁宅门说

入正门，一面砖砌大影壁上镌刻"迎祥"二字，背面有一篇330字的《大宅门说》，曰：

京华胜地，左环沧海，右控太行，南襟河济，北枕居庸，苏秦所谓"天府百二之国"，杜牧所谓"不得不可为王之地"，其势拔地以峥嵘，其气摩空而浩荡。大宅门在焉，乃其坐落。

麒麟仁兽，设武备而不为害，不履生虫，不折生草，以其祥瑞，乃其镇宅之宝。

父慈子孝，夫礼妻贤，兄友弟恭，达则治国平天下，隐则修身以齐家，是为大宅门之家风。

宅门之宾，腹纳挥天喝地之气，胸存润珠被玉之心，可以调素琴，阅金经，巧化鱼龙深潜水，描写飞天洒丹青。

厅堂轩榭，故都风貌；朱檐碧瓦，明幽曲折。才经琼岛春荫，忽现荷露垂虹，方历太液秋风，又见西山晴雪。燕京八景，四时纷呈。

琴曲茶饭，抑扬浓淡。品宫廷宗传，尝苏杭肴珍，赏南海生猛，探天府幽胜。忽闻二黄西皮，丝竹弹唱，生旦净末，京韵京腔，演绎市井百象，是为大宅门堂会之遗风。

温汤疗浴，内通六腑，外舒筋骨，去烦一寐琼台梦，无忧寝起赤子心。

无内之谓小，无外之谓大，宅门以大称之者，非君躬身亲验，何以知之？

⊙ 合院连廊

影壁之后的室内庭院，完全按照老四合院的布局，北为三间正厅，东西各三间带前廊的厢房，外廊吊挂楣子与坐凳栏杆一应俱全，正厅檐下悬有"雅趣"扇形匾额。

正厅之后会进入一个横长的过院，其北为大堂休息区；西侧有夹道，两边池水潺潺，分别通向客房居住区与康体休闲区；在两层空间中背墙均以雕花门窗装饰，且分别安装了不同造型的吊灯。

大堂西侧为西餐厅，砖砌旧式大门入口后，L形平面空间的墙壁上，你会发现主人精心保留下来的改造前后的许多影像资料。改造前后的天壤之别，令人感叹不已。

大堂东边为中餐厅，垂花门入口，恰好位于影壁东侧，跨入门内别有洞天：服务台面均做成传统雕花店面样式；大厅内安排八仙方桌及圆桌散座，东端设有一座戏台，两侧柱廊中分设八间格子间，楼上为十间半开敞的厅房；楼下厅西北有方亭水榭两座，悬"高山""流水""知音"匾额，更令座下水流清越，别有雅韵。

合院连廊的出入衔接，紧凑而巧妙，将宅院的不同功能区域划分得井井有条。

⊙ 畅音雅集戏楼台，格局配设巧思量

老北京的王府、贵宅、会馆中往往设有戏楼、戏台，如现存的紫禁城的畅音阁、恭王府的怡神所等均为代表。健壹宅院里的这座戏台，在有限空间中，描金绘彩，明艳夺目，锣鼓一响，粉墨登场，台上廊下，戏里戏外，如梦如幻。同时还呼应了厅前台下那"高山流水觅知音"的雅集之韵。

沿宽阔的楼梯拾级而上，一层平台悬挂的巨幅松鹤图，二层平台敬请的佛龛，下临长案和五扇雕屏，虽是过道空间，却同样引人驻足，耐人品味。

二、三层的餐饮包间或宏敞大气，或小巧别致。无论是大红漆门上的雕花门罩，还是恒信堂的《陋室铭》浮雕，抑或宏礼厅屋顶的精致宫灯，都精心布置，与房间格局和家具巧妙搭配。三层则采用对称格局，中厅开阔，前设坐床，后置书案棋桌，两翼四隔间：琼阴居、荷露堂、太液轩、晴雪厅，分别与昔日燕京八景中的"琼岛春阴""荷露垂虹""太液秋风""西山晴雪"一一对应，且含春夏秋冬四季之意。

二至六层的客房区，平面大致呈"回"字型环廊布置。房间依大小景观等不同，样式陈设装修各有特色。三层部分客房外临屋顶平台，借鉴日本枯山水铺砌卵石，成为静谧的小庭院。空间有限，巧思无量。

六层中央宽敞空间，南砌壁炉，北列书架，东置博古架，西设坐榻。东部的一张长桌实际上是用老门板改造而来，门边包镶铁皮，斑驳厚重，很有味道。

另外，宅院内的牡丹砖雕，修旧如旧的残垣门扇，以及一层残障人士通道与月亮侧门、荷塘鱼池的衔接等，都体现了建造者的巧思设计，和对北京传统宅院文化的倾心又关照到现代人生活方式的呈现。

总体而言，健壹迎祥宅院的格局非常复杂，但安排井然有序，室内一系列的庭院和厅堂空间，一方面汲取了传统四合院方正对称的特色，另一方面又引入戏园和小庭院等不同形式，彼此融合，相得益彰。这座建筑处处体现了京韵，其中的正房、厢房、亭榭、戏台、垂花门以及门墩、影壁、勾栏等细节局部均为中规中矩的老北京建筑式样；室内的柜、橱、衣架、床、椅、桌案等家具大多为旧式，有几处还特别布置了旧时清朝贵族所常用的坐床；墙上的装饰多为京绣和杨柳青年画，餐饮厅房分别以京戏剧目、古代乐器和燕京八景为主题取名，进一步展现了老北京的风采。

这宅院更像是一座名副其实的北京文化博物馆，客人到此，往往四处游览拍照，乐而忘返。海外客人尤其喜欢这里的氛围，感觉在此找到了如今已难得一见的旧时北京的风情韵致。

From 2003 to 2016, Kang Jianyi and his team have created three buildings of different yet unique Chinese styles in various locations of China's ancient capital Beijing.

JE Changping Manor

In 2003, with the establishment of the JE brand, the team took an old building in Beijing's rural Changping District, and refashioned it into a distinctly Beijing-style manor, with functions such as accommodation, fine dining, and theater performances.

The Changping Manor eschews the limitations of old-fashioned residential structures, tactfully combining a quadrangle courtyard, an opera theater and related facilities into a single complex. Each space has a unique interior/exterior design based on its function, and with Beijing opera as the overarching theme, many elements of traditional northern architecture are added to the complex's design, for a distinct atmosphere.

The most classic of Beijing structures is the quadrangle courtyard. These residential buildings, depending on the original owner's wealth and status, come in a wide array of scales and designs, but the common features of a symmetrical layout, hierarchically arranged living chambers, simple yet stolid style and private interior yard make the quadrangle courtyard the most ideal living space according to traditional Beijing aesthetics.

JE Manor, located on the most bustling street of Beijing's Changping District, features an innovative redesign, which transforms an old building into a distinctly Beijing-style quadrangle courtyard, infusing the complex with ingenuity and rich cultural elements.

Overall, the Manor is ambitious in conception, yet simplistic in execution, based on a traditional symmetrical design of a quadrangle courtyard, while further drawing on other elements, such as opera theaters and classical gardens. All of the rooms, as well as the theater stage, festoon gate and so on, are constructed according to strict traditional aesthetics, and the art found on the walls consists of Beijing-style paintings and embroidery. Finally, the private dining quarters are named after Beijing opera pieces, traditional instruments and scenic destinations, adding the final touch to the Old Beijing theme.

JE Manor is very much a living museum of Beijing culture.

右图：昌平宅院管窥
The right: A focused view of Changping Manor

五间巍峨大门，规制比拟王府
The towering front entrance, its scale on par with a royal manor

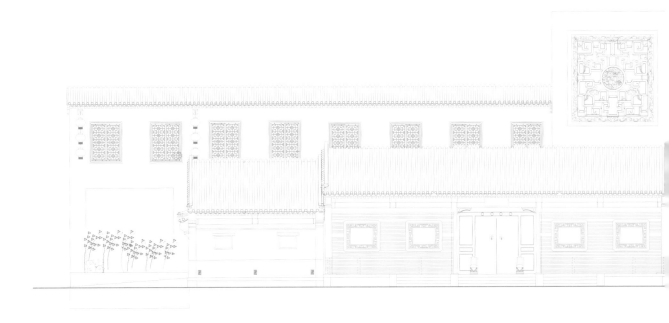

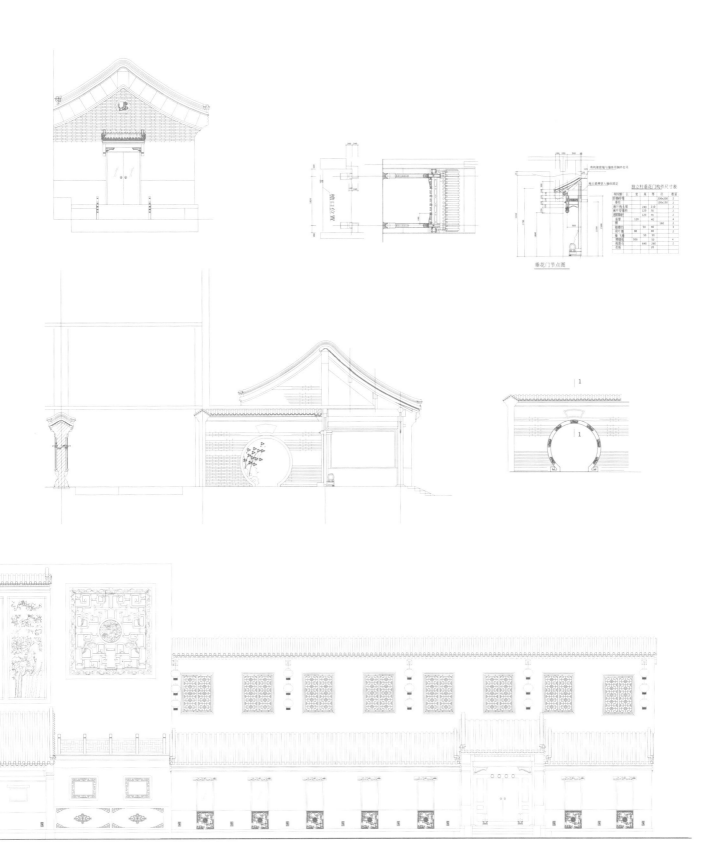

The Story of JE Mansion

赫赫内庭，奕奕华堂
The luxurious interior of the foyer

中轴对称的空间，内院厢房
The chambers are laid out symmetrically along a central axis

由大门、影壁、垂花门、正房、厢房围合而成的室内四合院
An indoor quadrangle courtyard, complete with entrance gate, carved screen wall and festoon gate

修旧如旧的残垣门扇
An antique-style doorway

大红漆门与室内陈设巧妙搭配
A red lacquered door integrates ingeniously with the other furnishings

无论是大门口镇宅的金狮，还是广亮大门前的拴马桩，都是大宅门不同的气派！
Whether it's a golden lion safeguarding an entrance gate, or a horse hitching post outside of a brightly lit foyer, each adds its own unique appeal to the mansion

右图：墙上旧砖雕，一篇古家训
The right: An ancient stone carving on the wall, bearing classical family teachings

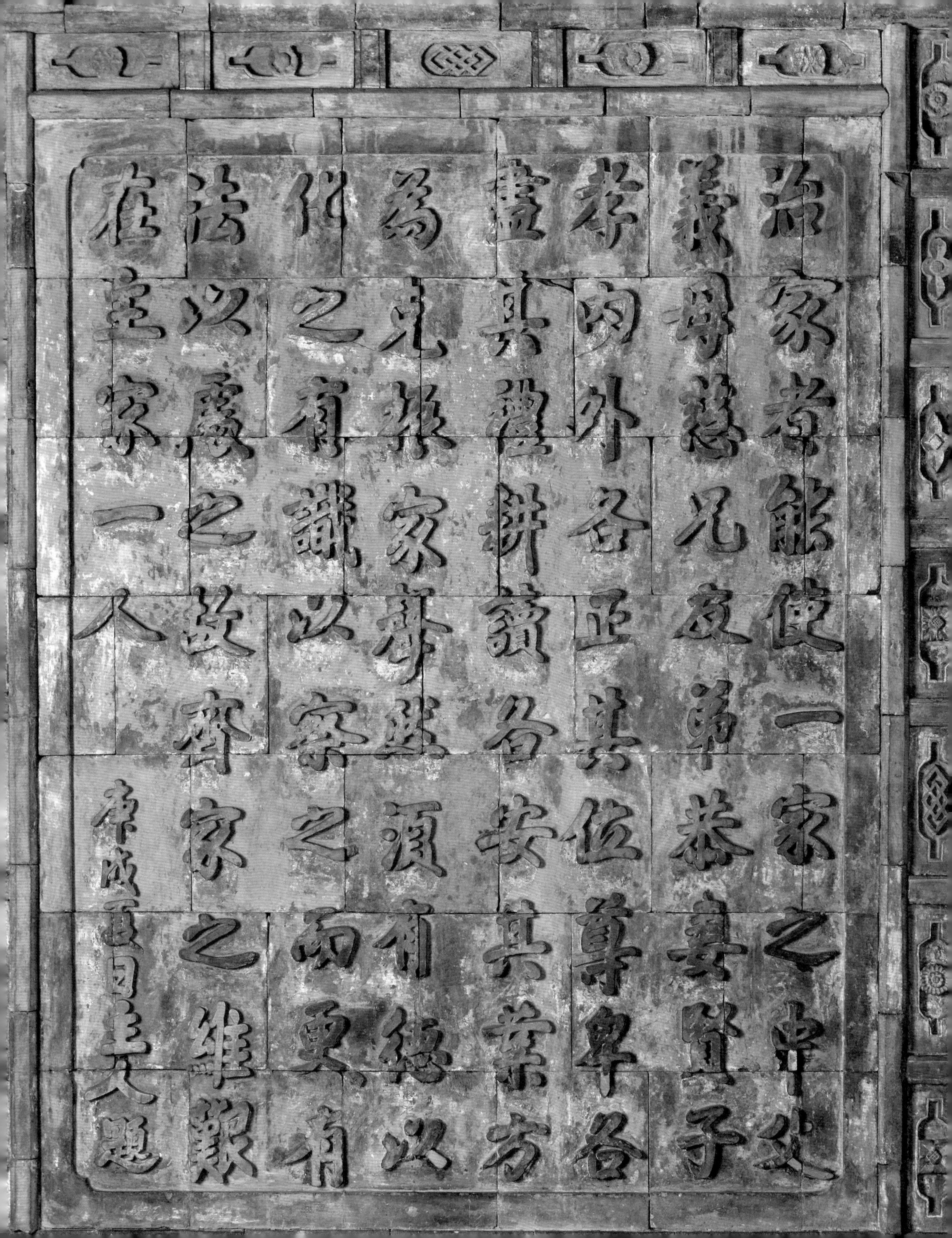
治家者能使一家之中父
義母慈兄友弟恭妻賢子
孝內外各正其位尊卑各
盡其禮耕讀各安其業方
爲克振家聲然須有總以
代之有識以寧之而更有
法以慶之故齊家之維艱
在主家一人
庚戌夏日主人題

旧门扇变身书案，宛如时光写就的简册
An antique door finds a new life as a writing desk

The Story of JE Mansion

卧室漏窗，透出几许温馨
The latticed window in this bedroom adds to the quaint environment

古旧家具带来悠远的历史味道
The old-fashioned furnishings add a historical touch

◎筑（著）者说：昌平宅院，是品牌初创时期的第一个项目，把收集起来的好东西归整并用传统的方式表现的同时，合理改造旧有格局是重点；现在回望还是有不少遗憾之处，但作为初心的起点，这个作品给了我和团队不少信心，引发我们继续在传统与现代融合的道路上探索前行。

⊙The architect/author says: The Changping Manor is the first project of JE. It emphasizes the reorganization of old-fashioned layouts, while displaying antiques in a traditional manner. Although the design is admittedly not without fault, as a starting point for our team, this project gave us the confidence to carry on with our vision of exploring the fusion of old and new.

宅院里的戏台描金绘彩，明艳夺目，凭栏赏曲，旧时京韵
The dazzlingly decorated theater is a great place to enjoy Beijing opera performance

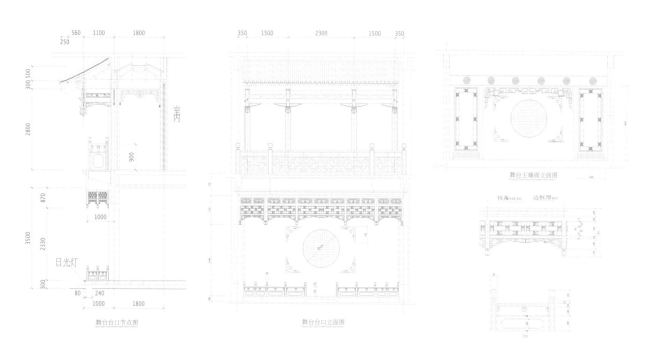

The Story of JE Mansion

朝阳静颐　健壹公馆

都市化发展使得北京迈入国际化、现代化的同时，也带来许多问题，诸如空间局促，喧嚣纷扰的紧张逼仄，尤其是中心商务区等地带。人们渴望案牍劳形之余无须舟车劳顿即可在市区内找到休憩小酌的放松休闲之处，健壹公馆正是在这样的市场需求之下诞生的。公馆位于朝阳区东四环，拥有大片草坪和复合楼层空间。相比健壹宅院而言，这座建筑更像是个人民间博物馆与私家花园的综合体，而且在中式传统建造风格之中融入了更多的近现代西方设计元素和技术手法，借民国新文化词汇，谓之"健壹公馆"，十分贴切恰当。

因受所在区域的限高规定约束，公馆整座建筑地上仅有两层，平面呈270度广角弧形展开，局部凹凸起伏，宛若锯齿。因为所处地形原为高尔夫练习场，本身高差明显，建筑的北立面全部凸起于地上，朝向宽阔草坪缓坡果岭；而南立面只有二层部分居于地上，首层与下沉庭院结合，掩藏在地平线以下。建筑外立面均以灰砖砌筑，大面积玻璃窗之上或设砖雕窗罩，或镌刻窗楣图案，除充分考虑室内采光功能外，室外楼体外观也不同寻常、精美耐看。

公馆整体为一花园院落，大门简洁低调，紧邻四环主路一匝道出口。入门之后，沿着一条蜿蜒旖旎、碧桃翠竹相间的绿荫夹道徐徐前行，便可来到主入口。南门前设门廊，三间悬山卷棚建筑，前设两座石雕白象，两侧坡道设有汉白玉栏杆，北墙中央启门，两边刻有《健壹公馆说》，文采与昌平宅院的《大宅门说》如出一辙。两侧坡道栏杆与楼体之间有水池喷泉，叮咚淙淙，是对光临客人最安静而又殷勤的欢迎。

门廊以三间纵向游廊联通门厅，廊间过道两侧有金鱼游弋、萍藻漂浮的石鱼缸做伴；门厅为三间歇山建筑，大门四枚门簪上书有"龙翔凤舞"，大门两旁石柱

外侧镌刻"周敦德寿，汉砖吉祥"八字隶书楹联，内侧为"鹿鸣山水静，梅放灵云香"楹联，室内东西两壁安设凤凰木制圆雕，古风雅韵扑面而来。

门厅北为二层大堂，中央位置设有一个八角形天井，与首层大厅贯通，天井之上建有一个方形平面的采光亭，空透明亮。

门厅北的迎宾前台，被设计成传统戏台样式，垂花柱、雕花挂落，引人入胜的礼宾迎客之道，别致而隆重。前台东西两层休息区，混搭而有序地放置了古式坐床、现代沙发和仿古旧式木椅；两侧墙上挂龙飞凤舞的绣袍作为装饰，与戏台样式的前台不动声色地搭配呼应。而大堂北面是视野广阔的宽大阳台，人们在此可临瞰茵茵苍翠的起伏草坪，绿云漫卷。

大堂东侧的精品屋中，也许你可以找到属于自己的那一份心头好。旁边宽大木制楼梯，拾级而上，便来到首层客房区。西面设有壁龛、石雕灯座，中间墙面隐刻唐代诗人杜牧《张好好诗》手迹；安静私密的小憩之所，以古典浪漫的唐诗来开篇。其间凹为餐聚场所的前厅，壁上素色麒麟砖雕，两侧有佳联相对："江畔轻云扶月出，雨余活水养花来"；前设长案，两旁为隔扇门，色调古朴。雅集与私密，古韵和新律，都市喧嚣中的一片"典雅绿洲"就这样铺陈开来。

餐饮区既设单独厅房，又有格间和大厅散座。独立厅房门扇上均绘有描金饰纹，匾额嵌于门侧。每一间的布局各具特色，房内墙悬与厅房名称呼应成趣的国画和对联；雅聚小酌之余，还可吟诗赏画，考证楹联的出处，仿佛又回到了旧时国人那种充满闲适雅趣的日子；按照传统格局，厅房室内多布置圆桌、古典座椅及旧式衣架，配饰和谐。

西端楼梯通向首层餐饮区，厨房内墙以现代手法处理，安装玻璃隔扇，开放通透，纤尘不染，窗台、灶台上还摆有盆花。人们移步路过厨房时，厨间内里设施，

司案掌勺的师傅，佳肴制作的流程，都可以一览无余。这处通透厨房，居然成为深受公馆客人喜爱，驻足一探舌尖美味出处的独特景致之地。

餐饮格间布置简洁，气氛幽静，临窗可观草坪院景；散座大厅分设多座八仙方桌或西式长条桌案、沙发，中央设吧台，墙壁上有雅俗共赏的老北京风情手绘。散座大厅之东还有一个过厅，通向四间独立的厅房。其中以合和堂入口为砖砌门楼形式，设有门墩，板门门扇附带辅首铁环，室内又分为前后两厅，以博古架分隔，西侧另设独立小书房，书卷气浓郁。启智堂可兼做会议厅，墙悬弘一大师手书《金刚经》和《心经》的书法珍品。尚义堂和恒信堂可兼做多功能厅举办会议，最为宽敞，东西墙上悬挂《太上感应篇》和《弟子规》。天长阁则是纯粹西派风格的长餐桌和高背椅，以及西式壁炉、酒柜。不同厅房，布置风格各异。

首层中厅通过天井与二层相通，厅内设有吧台和一张古典大圆桌。旁边另设独立西洋红酒雪茄厅，极为幽静，前厅设壁炉、酒柜，后厅设坐床、沙发，最适合在此会友、小憩。

客房居住区位于公馆建筑的东半部，首层设有休息厅，墙上挂着清代供奉宫廷的意大利传教士郎世宁所绘《功臣图像》以及画家任重所绘《高士图》。客房分设于弧形走廊两侧，上下共十八间，按照大小布局划分单间与套间，平面随宜布置。

建筑北面的草坪绿茵如毯，两边小丘缓坡起伏，颇有欧洲庄园气派。考虑到现代人们雅集聚会的需求，局部铺设木制平台，利用室外的木平台与绿草坪结合，布置露天桌椅和遮阳伞，绝对是举办私人庆典或品牌发布的不二场所。建筑的南侧设水池喷泉，纵向水流垂直悬挂在下沉庭院的墙壁上，宛如瀑布，带给端庄建筑几许生动流淌的活泼。

健壹公馆位于北京东部繁华的闹市区，总体建筑面积体量不算很大，功能也没有昌平宅院复杂，其空间相对集中，餐聚休闲和客房住宿区域左右对称，而在建筑和配饰细节上有很多精微的处理，内外空间浑然一体，尽显公馆的高尚气派与文雅氛围。公馆建筑内外，砖、木、石、雕、饰品等元素的镶嵌、搭配、陈列，可谓庭景交融，匠造陈设，相得益彰。

JE Chaoyang Mansion

While Beijing is becoming increasingly modernized and globalized, other issues have arisen, such as space limitations and noise, especially in the central and commercial areas, thus residents naturally seek a place among the cacophonies where they may find rest and respite. JE Mansion was created out of such market need. The mansion was adapted from a small-sized driving range, of which much of the lawn was retained. In comparison to the JE Changping Manor, this complex is even more like a cultural museum and personal garden, and the traditional Chinese buildings are infused with many contemporary elements and techniques.

The two-floor mansion features a 270 degree arc-shaped floor plan, with several points jutting out along the edge, like the teeth of a saw. The hills from the driving range remain, so that to the north side of the grounds is high and the south side low; also as a result of this layout, only two floors of the mansion appear above ground, the rest concealed below. The exterior walls are built of grey bricks, the large window frames decorated with various styles of carvings and engravings, providing both natural interior illumination and a unique aesthetic appearance.

The overall design of the mansion is based on a garden courtyard. After passing through the simplistic main gate, a winding path brings visitors to the entrance, which is covered by an overhanging gable roof. White marble railings line each side of the walkway, accompanied by a pair of white stone elephants. The walkway is split into two paths, between which a fountain bubbles quietly as it welcomes visitors.

JE Chaoyang Mansion is huddled away near one of the busiest sections of Beijing, and although it is by no means especially large in scale, and its functionality is less complex than JE Changping Manor, it is designed so as to maximize use of space, integrating living quarters and entertainment facilities into a single complex. These are further accented by brick, wood, stone carvings and various cultural artifacts.

大自然编织的花毯美妙迷人
Beautiful flower beds adorn the Mansion's grounds

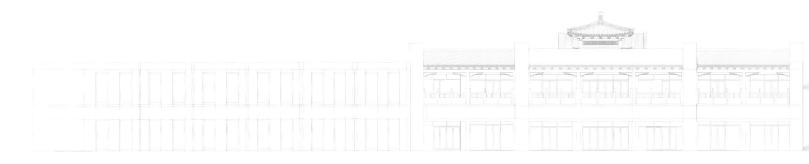

The Story of JE Mansion

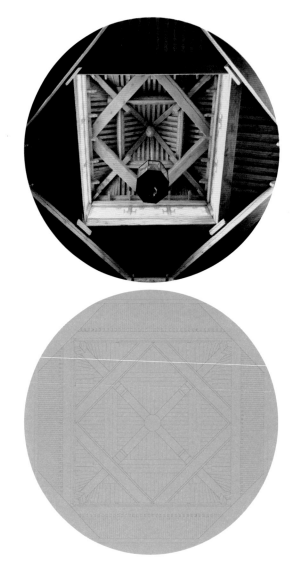

中庭天井贯通上下三层,八角形与方亭相互嵌套,
类似古代藻井的变体
The central skylight reaches up through three levels,
featuring an octagonal design based on traditional
sunken panels

右图:月亮门中的小庭院,宛如宋人扇面小品
The right: This courtyard, when viewed through the
moon door, resembles a Song dynasty landscape
painting

内廊墙上的刺绣,尽显高尚气派与文雅氛围
The works of embroidery hung on the walls of the corridors add to the elegant atmosphere

左图:绿色灌木与大片灰砖墙相互衬托
The left: The green shrubberies and large surfaces of the grey brick walls complement each other perfectly

The Story of JE Mansion

◎筑（著）者说：健壹公馆的修建历时三年，其间也故事不少，包括自己对自己的推翻和重新再来。比如前厅类似藻井的采光亭，原本是想沿用传统的藻井或亭的式样，但考虑到地上空间与下沉庭院的采光，几番尝试和调整后，采取了中西合璧的处理手法，所谓"不破不立"，在我们的作品中算是一例。繁忙城市中心地带要营造出一处清新雅静、空间容余又功能紧凑的会馆氛围场所，是整体的方向和目标；结构设计和布局配饰是主要功课。还希望带给朋友们别致的体验和感受。

⊙The architect/author says: The construction of JE Mansion lasted three years, and led to many stories. One of these concerns the skylight in the foyer, for which I originally intended to use a traditional sunken panel or pavilion design. Since I couldn't get the sunlight to extend all the way to the lower level, I ended up scrapping my design and rebuilding it with a fusion of Chinese and Western techniques. As they say, "There is no construction without deconstruction". My overall goal was to build a space that could offer respite from the busy city, while also providing a high degree of functionality, and my main task was to create a unique architectural and cultural experience for visitors.

右图：中西混搭家居配饰
The right: One of the rooms, outfitted with both Chinese and Western furnishings

远黛亭堂　西山景园

北京是一座拥有三千多年历史的古城，其西北郊一带山水尤为佳胜，金、元、明、清各朝屡屡在此营建苑囿、佛寺、道观，逐步形成一些名胜风景区；清代尤其对西山周围的海淀大加经营，修建了若干大型御苑，号称"三山五园"，包括万寿山清漪园、玉泉山静明园、香山静宜园以及圆明园、畅春园，还有乐善园、泉宗寺等其他次要的行宫和许多王公贵族的赐园，彼此连环相接，汇聚为壮观的园林建筑群，作为上层统治者主要的生活、理政之地。

健壹景园坐落在玉泉山西南，占地约三百亩；凭借其得天独厚的依山傍水（泉）的地理位置，以中式为主、西学为体的巧妙设计，精工构筑一系列湖、苑、亭、桥、厅、堂、楼、阁，优雅而有序地分布在视野开阔的私家园林之中，使人们置身山水间，享受到一处"游、居、行、望、思"的远黛庭堂，人们流连其中，时时会有"生活在别处"的感叹和赞许。景园的体量布局、优雅景致足以与昔日皇家园林相提并照，而被称赞为京郊西北经典园林胜地的一个新成员。

景园位于万安东路北侧，进入景园端庄简朴的大门后，沿着一条曲径环绕数百米的碧桃树夹道，方可来到主园区。春日桃花烂熳，热情迎宾，东晋陶渊明《桃花源记》描写武陵渔人偶然发现桃源深处古村的场景仿佛重现，"忽逢桃花林，夹岸数百步，中无杂树，芳草鲜美，落英缤纷……复行数十步，豁然开朗"。人们步行或驱车进入健壹景园时，往往会有偶遇桃花源的惊喜和感受。

景园这一现代桃花源，既承古典历史之韵，又添现代怡然之律；园区视野平旷，厅堂屋舍俨然，有良林美池，水榭楼台错落，小桥曲径蜿蜒，移步换景，赏心悦目。

一条溪流环绕园中，路旁水声清越，可见一座水心榭端立于池塘中央。这座名曰"康宁阁"的亭阁，纵横三间方形平面，重檐攒尖顶，四面各出一间悬山抱厦，周围设汉白玉栏杆，以曲桥连接池岸。门窗采用双交四椀菱花，中央为正方形大厅，四面抱厦被分别设为门厅、备餐间、洗手间和休息厅，北墙悬挂一幅"福禄

健壹景园全景图
Panoramic design of the JE Garden

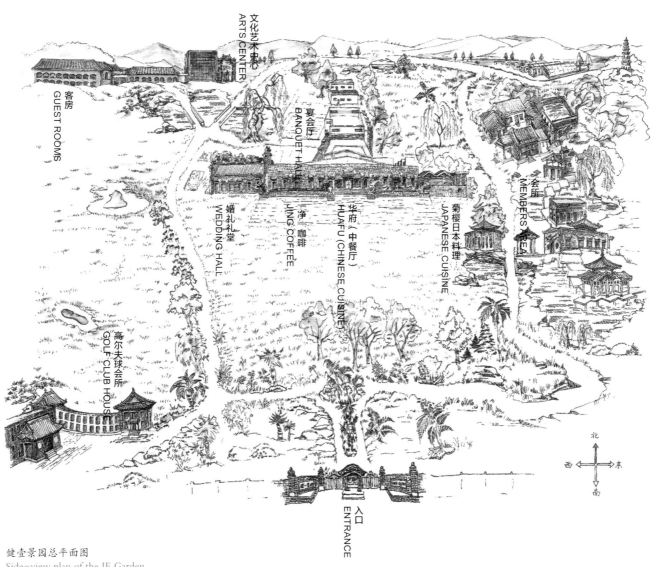

健壹景园总平面图
Side-view plan of the JE Garden

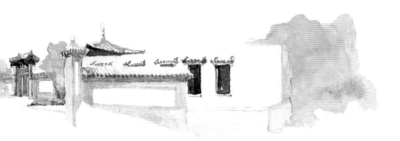

寿"古画，南面墙上则是著名学者梁启超所书"且欲近寻彭泽宰，不知谁是谪仙人"，既应景又衬托出人们在此雅集时的逍遥氛围。

在水心建殿堂、楼阁或水榭是中国古典园林的常用手法，北宋张择端《金明池夺标图》就描绘过东京御苑中的水心殿形象，清代颐和园中的治镜阁和绮春园中的鉴碧亭都是类似的景致。康宁阁的造型原拟模仿紫禁城御花园的千秋亭和万春亭，后因考虑到圆形平面施工难度和整体景园的体量而改为现在的形式。

经过康宁阁，主道上的一座牌楼作为园子的第二道大门，一条笔直的道路南北贯通，东西两侧均设健壹书苑，包含二十间餐聚雅集的独立包间。书苑东院正门朝西，前设三间硬山卷棚顶门廊，明间柱上悬挂楹联"福寿康健，天行健，健而康，厚德载物，物华天宝；万法归壹，道生壹，壹生二，三生万物，物外超然"，上下联中嵌有"健壹"二字，彰示品牌于含蓄书法之中。入口两侧为雕花漏窗，门上嵌有"双峰挺秀"旧石额匾。入门穿过一间廊子入二门，即来到东院大堂，由一座古民居迁建而成。

大堂东为内院，院中以细砂铺地，精心耙梳出水纹，砂上点缀小巧石块和精致石灯，为典型的枯山水庭园。院北健壹堂是景园内最大包间，纵横三间方形平面，攒尖屋顶，南北各出抱厦一间，室内通高达14米，东壁为东晋王羲之《十七帖》的书法浮雕，西面悬斗大"福""寿"书法墨宝。阳光从四面高窗倾泻而下，明亮畅透，一如主人的秉性和做派。

内院南北两厢的延禧宫和清心居，均采用金龙雕花大门，室内以传统的飞罩隔断分隔空间，通过大窗可以看到南面的康宁阁或北面佛堂一带的风景。大堂两侧各有一个长条形水院，院侧为城铭轩和馨月斋，采用石框红板门，门簪上各自题写"福寿康宁"和"人寿年丰"，充满吉祥祝福寓意。门内设有前厅，厅旁放置传统坐床，宽阔的厅内除合围餐桌和座椅、柜子之外，还设有一架旧式大床，最宜于喜庆婚宴。

从中庭楼梯拾级而下，可至地下层，中央设有开敞的吧台、散座区和部分明厨房；两侧分别布置琼阴居、荷露堂、太液轩和晴雪厅四个包间，走廊巧妙利用相对幽暗的地下空间设计布置灯光，翠竹掩映于漏窗之中，鸟笼式的吊灯更增添了闲雅气氛。包间布置都是中西合璧风格，外窗正对下沉庭院，光线柔和。

书苑西院与东院相对，采用垂花门形式，门上悬"书为善宝"匾额，两侧立石雕白象。入门是一个卵石铺地的前院，西为大堂所在，方形平面，木结构攒尖顶，门内侧左右墙壁上分别镌刻《心经》和《吉祥经》。大堂格局对称，两边各设旧式坐床，壁上有京绣展示点缀。东侧走廊串联五间包房，外窗正对西面的缓坡小丘。从大堂西侧的木楼梯而下，可至西区地下层，也设有五个包间，格局大小互异，外窗正对水流涟涟、池草素雅的下沉庭院。

从东西两院的地下层均可穿越砖门进入酒窖连廊。酒窖大厅为椭圆形，中央设品酒长桌，两侧密置酒柜，宛如欧洲城堡的豪华密室，等待品鉴者来开启佳酿的密码。

右图：婚礼堂外墙与池水相互衬托
The right: An exterior wall of the wedding hall and the water running next to it serve as relief to one another

沿南北向的大道继续北行，穿过河上长桥，可来到后院，院南立有大型太湖石一尊，造型奇秀；院北屏风后另立一块造型敦厚的巨石，以作呼应。

景园内建有一处两进院的佛堂"菩提阁"。佛堂前建三间山门，门前地势开阔，入门为前院，过垂花门，可来到游廊环抱的主院，院中设水池，池上石雕佛龛。院北为五间正殿，两侧东西配殿各三间，内部均表现为藏传佛教风格。北京西郊皇家园林中建有多座佛寺，且以藏传佛教殿堂为主，健壹景园中的这座佛堂继承了这一文化传统，还专门礼请了两位来自西藏的喇嘛在此修行，赋予整个园区以吉祥宁静的超尘氛围。

景园西部为宴会区，西向入口门前有大片草坪花坛和素石喷泉，两侧竖立十二根石柱，柱头分别雕刻十二生肖图案。西向大门上悬"健壹景园"匾额，门厅中央竖立一尊《千里走单骑》抽象雕塑，访客光临，先来会一会古典文学作品中的先贤。门厅之东为西式宴会礼堂，宽20米，长30米，可容四百余座位，旁边还附设休息厅。门厅北为宽阔的过厅，其西侧布置精品屋、休闲走廊和椭圆形会议厅；二楼设有宏礼、崇仁、启智、尚义四间多功能厅，可兼作会议和宴会使用。私密雅集、开放敞阔兼顾。

宴会区以南称为华府，是以一道月亮门与过厅分隔的散台雅座区域，玻璃屋顶下，事事如意的柿树与窈窕旖旎的修竹，将华府布局得井井有条，又动静兼顾，三五知己聚餐小酌，或设宴立围宴请嘉宾，都可安排得恰如其分。西南翼的"净咖啡"，临窗享受绿草茵茵与池水潺潺，或是于室外藤沙发上享受午后的阳光，都是下午茶时光的最佳去处。

华府之东设有一座独立的中式礼堂，主要用作婚礼庆典。外观其筑，南北两面为徽式马头墙，筑内中央部分为九开间大厅，内立双排共二十根内柱，寓意"十全十美"。两侧大落地窗带来明亮欢畅的效果。礼堂东端还布置有一顶旧式花轿，描金织红，喜庆隆重，愈加烘托了这喜气长存的氛围。

华府宴会区往东，是景园的客房居住区；在南北大道的西侧设有日本鸟居式入口，门外以沙石铺就一片银滩，旁立石雕落地座灯和主人收藏的仕女石像，既有照明功能，又做装饰欣赏。居住区设有中西合璧风格的大堂，内里结构是一座从浙江整体搬迁而来的木质古民居；旧式廊柱牌匾、楼栏台阶，有机融合在现代外延砖砌楼体之中，时光在此交汇，道不尽遐思无限。

从大堂西行，穿过北侧一个枯山水庭园，可进入一条地板架空于砂地之上的木结构长廊，两侧为十三间客房。客房套型各异，但都富有简洁雅致的情趣，有些客房还带有院落，小院中绿植山水，景致宜人。客房居住区西侧的玻璃走廊，连通了华府宴会区。景园里所有的室内建筑部分，均由地上或下沉廊桥汇通。

相比于昌平的宅院和朝阳的公馆，健壹景园是一座完整意义上的私人园林。其中宴会区由旧建筑改造而来，其他设施均为新建。受所在地区限高的影响，绝大多数地面建筑仅有一层，但内部空间变化多端，外部造型错落起伏，体现出丰富的建筑群体效果。周围环境充分与缓坡小丘、草坪曲径、水系山石结合，并局部设置下沉庭院，充分展现了中国古典园林特有的曲径通幽、移步换景之美；更借景玉泉山玉峰塔，形成了京西郊新园景。

JE West Mountain Garden

Beijing is an ancient city with over 3000 years of history, and its northwestern outskirts are particularly rich in culture and scenery. Throughout the Jin, Yuan, Ming and Qing dynasties, many temples and monasteries were built there, followed by manors and temporary palaces built by royalty in later Qing times. The structures now form an impressive collection of different styles.

JE Garden is located on the southwestern face of Yuquan Mountain, and covers an area of 50 acres. The complex features a mainly Chinese design, with Western elements added, and includes components such as lakes, rivers, bridges and towers, arranged in a stunning private garden assembly, so that visitors may enjoy the scenery and tranquility as they stay there. The complex's elegant layout is highly reminiscent of an ancient royal garden.

In comparison to JE Manor and JE Mansion, this instalment in the series is much more like a traditional Chinese garden complex. The dining section was constructed from an old building, while the other structures are all new. Due to local building height regulations, most of the complex is only one storey high, yet the interior designs of the spaces vary widely, and this shows on the outside. Throughout the garden grounds are rolling hills, lawns, walking paths, streams and ornate rocks. These, along with the Yuquan Mountain Pagoda visible in the distance, combine to form an impressive scene.

浓墨重彩的外立面
Colorful outer wall

The Story of JE Mansion

85

景园主体建筑和草坪,掩映在垂柳繁花之间
The main building of the Garden and its lawn, set off among weeping willows and blooming blossoms

碧波倒影康宁阁，四面清澄水心榭
Tranquil water surrounding Kangning Pavilion

The Story of JE Mansion

夕阳下的礼堂正面,高高的马头墙,双扇木板门,大气庄严的景园礼堂
Front view of hall at sunset. The high horse head walls and double wooden doors give the structure an air of solemnity

雪后菩提阁的澄净世界
The Bodhi Monastery after snowfall

左图：晶莹世界，树影婆娑，楼台起伏，宛若仙境
The left: Shadows of trees on the snow covered ground, the cascading complex seen in the background

The Story of JE Mansion

93

宴会厅现代雕塑，生旦净丑，粉墨登场
Modern sculptures of the four Beijing opera roles, found in the main dining area

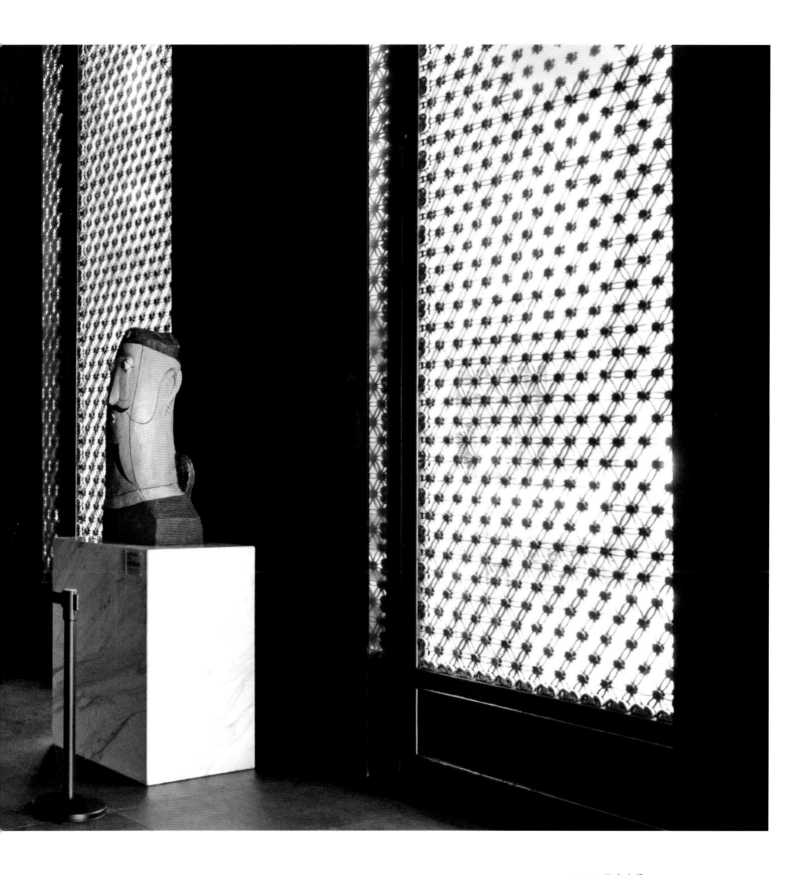

下页图：礼堂内景
The next page: Interior of the hall

老匾额下的古朴坐床
The traditional-style bed under an antique wooden tablet

左图：康宁阁内木构空间
The left: Interior wooden structure of Kangning Pavilion

对称严谨的空间，蕴含传统的礼仪秩序
The symmetrical layout maximizes space and adds a sense of orderliness

The Story of JE Mansion

江南厅堂中的火炉,照亮历史的记忆
The southern-style lamps bring visitors back in time

下页图：健壹堂红色的福寿斗方墨宝成为凝重木色中的点睛之笔

The next page: Works of calligraphy on red paper hanging on the walls add the final touch to the hall's rustic feel

The Story of JE Mansion

棋桌一局手谈，胜读二十四史
A chess table in the center completes the room's traditional theme

雕花金龙大门
Carved "golden dragon" doors

◎筑（著）者说：景园是品牌建立近十年时较为完整的一个作品，也是我和团队一路探索实践的一个小结。因为项目空间更为广阔，湖、苑、塔、亭、桥、坪、廊、阁齐备，需要考量关照的要素更多，特别是要考虑自然山水、人文景观的和谐共存，这也是我们修建历时最久的一个项目。我和我的团队很幸运也很骄傲，可以在历史悠久的北京西山风景区，完成这样一个中式为体、西学为用、有继承、有创新的建筑作品。其中有些局部，更是我个人建筑设计理想的小小实现，比如康宁阁，就是比照自己心目中的木作极致金阁寺所建；比如菩提阁，则为心中的信仰投影到现实的波心福田。

⊙The architect/author says: JE Garden is one of the most complete works we've done in the decade or so since our founding, and a conclusion to one of our routes of exploration. We had a lot of space to work with, the grounds including lakes, lawns, bridges, corridors and lots of other things, but since we had to be aware of protecting the natural and cultural environment there, the project took longer than any other. My team and I are very fortunate and glad to have been able to complete this innovative complex within the scenic area of Beijing's West Mountain area. Many of the structures were dreams come true for us to make, like the all-wooden Kangning Pavilion.

篆书"寿"字纹黄色地毯
Seal script characters adorn a golden carpet

旧式婚床，雕饰华美，龙凤呈祥
Old-fashioned wedding bed, with ornate furnishings

3

工筑文法要义篇

Styles and Techniques

健壹营造的建筑，给人第一印象是浓郁的中式风格，而中国建筑在世界建筑群体中，可谓自成一体；中国建筑的历史与中国的文明文化相伴相生，源远流长。中国人一向采用的本土营造体系和设计构思，存在着一些从古至今未曾改变的经典特征，如同语言的文法，中式建筑也有着它独特而经典的文法要义。而这些特征，即使在推陈出新、充满创意的健壹作品中，也被栩栩如生地传承了下来。

让我们从如下几项中式建筑之独到文法（grammar）来品鉴。

大木构架

中国古代建筑最大的特色就是以木头为主要的建构材料。"木"是中国五行元素之一，也是所有常用建筑材料中唯一来自生物体的原料，虽然强度和耐久性不及砖石，却具有运输方便、容易加工等优点，同时也有很好的抗震性能，只要精心设计、注意保养，用木材同样可以创造出雄伟坚固的建筑，并长期屹立不倒。

典型的中国古代建筑中，根据木构件的作用分为大木作和小木作两个系统，前者指承重的结构部分，比如梁、柱、檩、斗拱，组成严密的框架体系；后者指围护、装修性的附属部分，比如门、窗、隔断、天花。木质材料天然具有温润的质感，同时经过精心的排列组合，表现出浪漫的线性之美，与书法艺术殊途同归。例如上下均有收分的棱柱和弯曲的月梁，在视觉上富有弹性的力量，屋顶椽架的举折之势更有一番灵动之气，被《诗经》赞誉为"如鸟斯革，如翚斯飞"。

健壹营造的工程或改建，或新建，或移建，情况比较复杂。移建部分本来就是古旧的木头老房子，其新建部分和改造中添加的部分也大多采用木结构，传承了中式建筑中木为首要元素的传统；而改建和新建部分的木

材选择，采用了保证原料材质量并适合北方气候的进口木材。

健壹营造的主要木构架属于典型的清代官式抬梁式体系，严格采用榫卯方式连接，其基本原理是：首先在台基上竖立柱子，然后在柱子顶部承托横向的梁，梁上再竖立一种名叫"瓜柱"的矮柱子。瓜柱上再承托梁，梁上再加瓜柱，就这样一层一层地抬上去，最上面的那层梁上竖立一根位置最高的脊瓜柱，由此构成一个坡屋顶的轮廓。然后在每层梁的头上分别搁上檩条，檩条上钉椽子，铺上望板，最后再铺上瓦，就完成了最重要的大木作构架系统。此外，柱子之间还需要用一种横向的枋子加以串联，加强柱子彼此之间的联系，形成更稳定的整体结构。

健壹建筑中有几座从江浙地区搬迁来而来的老房子，如景园的书苑前厅，客居大堂，则属于穿斗式的木结构，柱子相对偏细，以柱头直接承托檩条，柱间以细长的穿枋和斗枋加以串联，显得更加灵巧。

中式单体建筑的基本特征：垒抬提升的台基或柱座，木结构的立柱和横梁，层层垒高的同时横木长度向上依次缩短，至最短横木上架设中柱，而构成三角形的稳定结构，用以支持屋顶；每一副梁架，组对间隔排列，由梁架立柱顶端延伸到相邻的横梁或系梁，连接而组成"间"或"格"的空间；若干"间"和"梁架"组成矩形的建筑；横木各层末端架设屋顶常带坡度而有飞檐，这些都可在健壹的作品中得以亲见证实。

景园中式婚礼堂，贯通木质立柱加石墩柱基的设计与手法，即为最典型的示例。由木结构搭建而成的建筑类型包括厅堂、楼阁、亭、榭、游廊等，屋顶呈现歇山、悬山、硬山和攒尖等几种不同造型，也是中式经典的例证。

Upon visiting the structures designed by JE, many people's first impression was that the buildings were highly Chinese in style. Chinese architecture is closely linked to Chinese history and culture. The Chinese have continued to use their own techniques and concepts, so that Chinese structures have carried on a set of classic characteristics that have remained more or less unchanged, much like the basic pattern or grammar of a language. This architectural "grammar" is embraced by JE, and forms the basis upon which innovation is added to complete their projects.

Next let's take a look at the unique "grammar" of Chinese architecture.

Wooden Framework

The most distinguishing element of classical Chinese architecture is the use of wood as the main material. Wood is one of the five elementals in Daoism, and is the only commonly used building material to derive directly from nature. Although not as durable as brick, wood is easy to transport and process, and possesses natural anti-vibration qualities. Only with careful design and maintenance may large and impressive structures be made of wood. In Chinese architecture, large pieces of wood are reserved for complex systems of beams and pillars, while smaller pieces are used for components like doors, windows and ceiling panels.

The wooden frameworks used by JE are based on a classic Qing palace-style raised beam network, known as the "mortise and tenon joint" system. First the foundation is built, upon which pillars are raised, then beams are laid to connect the pillars, with short columns protruding from the tops of the beams. Next, more beams are laid, and more short columns are added, creating a latticework one layer at a time, the final layer forming the outline of the room's ceiling. Next a layer of purlins is applied to each layer of beams, to which rafters are attached, and these are in turn covered with sheaths and tiles. This framework is then further strengthened by a series of horizontally placed wooden blocks.

The basic characteristics of single Chinese-style buildings are a raised foundation (or column base), wooden pillars and beams, then each layer of the beams becomes continually shorter, with the smallest pieces at the top arranged in a sturdy triangle formation, to support the ceiling. Each beam is spaced at intervals, in a radial arrangement, forming an overall square-shaped structure. The ends of each of the beams are sloped and upturned. All of these elements can be seen in JE's works.

The JE Garden's Chinese-style wedding hall is a perfect example of this system, featuring a stone foundation and wooden pillars. Other wooden structures built according to this design include pavilions, towers and covered walkways. Different types of ornate roofs can be constructed atop the latticework, each uniquely Chinese in style.

经过精心排列组合的木柱表现出韵律之美
A melodious design created through careful woodwork arrangement

右图：江南古建筑的垂花柱与穿枋
The right: Southern Chinese-style decorative beam

檩条上布置椽子，上面再铺设望板和屋瓦
Purlins layered with rafters, in turn covered with sheaths and tiles

江南古建筑中形如弯月的月梁，尚存唐宋古风
A moon beam, a style found in the south, originating from the Tang and Song dynasties

江南古建筑的牛腿布满木雕，灵动鲜活
Vivid classical southern-style carvings adorn the corners between pillars and beams

檐柱上的雀替，起辅助支撑作用
Decorative "sparrow braces" attached to an eave column for support

所有木构件严格采用榫卯方式连接
All wooden frameworks adhere strictly to the mortise and tenon system

走廊的木构体系
Wooden framework of a covered walkway

木结构同样可以带来高大明畅的室内空间
A wooden framework can create a lofty and well-illuminated interior space

◎筑(著)者说：盖房子就好像是搭积木，只有充分把握木结构的所有特征和优缺点，才能赋予建筑以坚固而精巧的骨架系统。木结构是中国古典建筑的精髓和别致所在，在我们的作品中也是充分考虑和遵循的传统之一。而在新建和改建部分的取材方面，则要考虑就地取材，依旧修旧；另外也要考虑北方四季分明，寒暑温度、湿度变化大等。所以在新建部分，我们根据设计，引进了部分俄罗斯优质木材作为基础料材，这也是现代建造工程因条件允许，有了更多的备选方案。

⊙The architect/author says: When building a wooden house, only by having a firm grasp of all the characteristics, advantages and disadvantages can you create a framework that is clever enough in design to be sturdy. Wooden frameworks are an essential and unique element of classical Chinese architecture, and in our projects it is one of the main traditions that we do our best to respect and follow. We also strive to procure as much material as we can locally, but as northern China has four distinct seasons with varying temperatures and humidities, we imported some high-quality wood from Russia for part of the framework, which is an option that modern construction has made possible.

砖瓦土石

中国古代建筑虽以木结构为主，但砖、瓦、土、石也是重要的营造材料。这些材料在中国传统五行中均属于"土"的范畴，建造时与大木作、小木作密切配合，因此自古以来中国人又将建筑通俗地称作"土木工程"，可见这一元素在建筑整体过程中的重要和不可或缺。

砖，是用黏土烧制而成，中国古代很早就出现了砖，但以往产量较少，只在建筑的一些局部位置使用，明代以后才开始大规模使用砖，提高了建筑的坚固程度和保温、隔热效能。健壹营造大量使用了北京地区传统的青灰砖，其中砌墙用停泥砖，铺地用较大的方砖，影壁、檐口、博缝、戗檐等特殊部位使用特制的砖。槛墙、山墙下碱、影壁等最讲究的墙体用"干摆"的方式砌筑，也就是行里行外都耳熟能详的"磨砖对缝"，每皮砖之间不留砖缝，从外观来看砖块之间紧密地组合在一起，浑然一体。

瓦，北方地区常用的灰瓦，分为筒瓦和板瓦两类，板瓦上仰铺为底瓦，筒瓦下覆一块压一块地向上逐步安设。大多数坡屋顶的两个坡面相交的最高处被处理成圆弧的形式，称作"卷棚"，少数屋顶带有正脊、垂脊和各种小兽，包括吻兽、垂兽、仙人骑鸡、海马、狻猊等，也都属于瓦作范畴。健壹作品中凡采取中式设计的屋顶，都体现了这些传统瓦作特征。

特别端详一下菩提阁的屋顶和飞檐：菩提阁是私人园林中的佛堂建筑，屋顶既考虑整体的沉稳，又体现对供奉神佛的敬重；因此屋顶是由稍微倾斜的斜坡平面层层组成（通过调整梁架每一层梁的长度和高度而实现），在屋顶坡面两两相交的屋脊，通过加高的线脚予以强化，并以神兽饰物点缀；而飞檐的进深，也是设计重视的细节，夏季可以使整面南墙处于檐影遮蔽之下，以保证夏季的凉爽；而在冬季，又能充分采光，让阳光可以一直照到房屋后部。

健壹营造中石材用得很有节制和讲究：遵循古法，改建和新建部分，只在台基、门墩、门框等部位才使用结实耐磨的石料，栏杆和桥一般则采用汉白玉；为讲求创新，在公馆和景园的下沉庭院，以及景园礼堂外水景喷泉池塘的基底，则采用了黑色大理石，以体现水流清澈，也是考虑到日常清理的便利。

Brick, Tile, Clay and Stone

Although classical Chinese architecture used wood as its main building material, brick, tile, clay and stone are also frequently combined with wood. Traditionally these materials fall under the elemental of "earth".

Brick, made from firing clay, came into large-scale usage in China after the Ming dynasty. For JE's projects large numbers of traditional gray lime bricks procured from the Beijing region were used. Packed earth bricks were used for the walls, while larger ones form many of the floors, with specially made bricks for other components, such as screen walls and eaves. For many of these decorative elements, a method was used in which there appears to be little or no seam between each of the bricks.

As for tiles, in northern China gray tiles are most common, and include round tiles and panel tiles. Panel tiles are laid first to form a foundation, then the round tiles are laid so that part of each row covers the previous row. Most sloped roofs have a rounded spine down the center, and some feature small carvings of animals and mythological creatures. These carvings are part of tile work. JE's series of Chinese buildings all feature such finishings.

The roof and eaves of the Bodhi Monastery: As a religious structure, both stability and reverence to Buddha were considered in the roof's design. The roof has a gradual slope, and at the spine where the two sides meet, mouldings are placed for extra strength, with decorative animal carvings for ornamentation. The depth of the eaves was also an important detail, as in the summer the eaves can provide ample shade, while in the winter they can accommodate additional light.

景园礼堂的灰砖山墙
Gray brick wall of the garden's hall

汉白玉栏杆,由望柱、瘦项、华板、束腰组成,
比例和谐,雕琢细致
White marble railings, with pillars, decorative panels
and girdling, for a subtle yet refined look

The Story of JE Mansion

127

砖石结合，层层雕饰，磨砖对缝尚犹存
Brick and stone combined, each carved with ornate patterns

The Story of JE Mansion

昌平宅院墙上浮雕
Relief carving on the wall of Changping Manor

左图：菩提阁的石雕佛龛，宛如精致的小亭
The left: The carved stone altar of the Bodhi Monastery, resembling a small pavilion

圆形石雕，镶嵌在墙壁上，形如拱璧
A large round stone sculpture inlaid on a gray brick wall

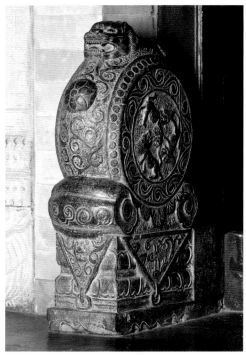

大门旁的门砧石，又名抱鼓石，俗称门墩
Stone sculptures on either side of the doorway

技艺高超的石雕
Displaying intricate stone carvings

宅院的标志石雕

Calligraphic carvings at the manor

◎筑（著）者说：考古学帮助我们了解到在中国先人们最早的居所，都是黄土地层垒墙而起；殷商时期的王宫都会有中大型夯土台基，均铺石块，覆青铜圆板后上置柱础；而后屋顶飞檐、花园亭台的发展，更少不了"土"元素部分对建筑的贡献。我们的三个建筑作品，均坐落在中国的北方地区，在秉承中式传统营造章法之外，我更着重强调的是：因地制宜、就地取材、节制讲究（毕竟项目的空间和建造条件都有限）；把握好度，可以令砖、瓦、土、石等在完成其建筑功能的基础上，也散发出它们各自的艺术之美，即它们是建筑材料，同时也是艺术品。当然，讲究里面还包括了工艺技巧，我们团队里工匠师傅的手工技艺达到"磨砖对缝"是基本标准。

⊙The architect/author says: During the Shang dynasty, royal palaces were built atop stone foundations, which were covered with bronze plates before the pillars were erected. Later, in the development of eaves and garden pavilions the element of "earth" was equally indispensible. All of our architectural projects are located in northern China, and aside from adhering to traditional building methods, we must emphasize local adaptability, usage of local materials and resource control. If used properly, brick, tile, earth and stone can effectively exhibit their dual function as building materials and works of art. Of course, this involves a wide range of techniques, and our craftsmen strive to achieve the standard of "carving the bricks so that they fit perfectly".

色彩漆画

中式建筑中色彩的装饰效果，据说最早是基于保护木材的目的而使用油漆材料，大约在一千多年前才被建筑师发现并应用到极致。根据《营造法式》的记载，及至宋朝，油漆的装饰性使用已成固定程序，如重大建筑的柱体、门窗和墙体，通常为朱红色，而上方门楣和斗拱装饰有以蓝绿色为主体，以朱红色和金黄色为点缀辅色的图案。这样一来，向阳的暖色调，和阴影部分的冷色调，加强了光影对比。

这些传统的色彩漆画手法，在健壹营造中既有传承，又有突破：大多数情况下，健壹营造的木头梁柱都没有油漆彩画，尽量保持原材料的素净本色。这一设计是基于对所在营建环境（空间有限、层高局限等）和现代审美（简洁、轻装、舒适）的均衡表现。但在某些特殊局部，色彩传统又被恰如其分地传承沿用，比如昌平宅院的戏台、西山景园净咖啡厅的外门，都特意以浓墨重彩处理，形成视觉焦点；朝阳公馆和西山景园等餐饮雅间的屋门也都是或朱红重漆或镂花洒金，以示隆重。

附着在木构件之上的精美木雕和油漆彩画也是中式古建的重要特征，为建筑添加了微妙的细节和绚丽的色彩。这一特色在健壹营造的移建古房屋和老物件方面体现较为充分，精心清理，复原功能后，尽量保持古屋和物件的原始色彩，让它们在现代的光影和今人的注视下，述说原本的故事，散发古典的气息。

Color and Lacquer

The usage of color in Chinese architecture originates from the protection of wood, and it was not until just over a thousand years ago that architects used paint to its full potential. According to historical records, as of the Song dynasty usage of color was relatively fixed. For example, for large structures, the pillars, doors and windows were usually crimson, while door lintels and ceiling brackets were green, with crimson and gold for ornamental patterning. This created a stark contrast between warm and cool colors.

These traditional color and lacquer techniques have been both carried on and innovated upon in JE's works. In most cases the wooden beams and pillars are left in their unadorned, natural state, in accord with the modern aesthetics of minimalism and sustainability. But in some places, traditional heavy coloring is used for certain effect, such as in the opera theater of the Changping Manor, the doors of the West Mountain garden's cafe, so as to make them visual highlights. The walls of most of the private dining rooms are also painted in crimson or stamped with gold foil, for an atmosphere of grandeur.

Exquisite carvings and colorings on wooden components are important characteristics of Chinese architecture, and add subtle detail to the beauty of the structure. When using antique structures and objects, JE strives to retain their original colors and designs, so that their stories may be carried on for modern society to enjoy.

门窗以朱漆镂花洒金来表现隆重的气息
The crimson doors and windows and gold ornamentation create a grand atmosphere

井口天花彩画戏台
The colorful theater with a raised audience level

古雅与浓重交织
Antique styling and rich coloring

◎筑(著)者说:色彩设计方面,我考虑更多的是,如何让古旧的部分继续固有的经典,让新鲜的部分明快均衡并匹配旧有;所以我们不追求张扬,而更在意如何从容地将古典和现代更好地融合在一个调色板上。就整体建筑而言,我们对素色和原色更注重,而经典的彩色和漆画,都是用于点睛之处。

⊙The architect/author says: When designing color plans, I focus on retaining the classical qualities of antique components, while harmoniously balancing these with new elements. So rather than merely displaying tradition, we're more interested in combine classic and contemporary on one palette. Overall, our works are mainly plain in color, with paint and lacquer used for the finishing touch.

The Story of JE Mansion

内外格局

健壹营造的三处建筑，有单体，也有组群；有改造，也有新建；有合围封闭，也有敞开立体空间；有独立院落（西山景园的菩提阁），也有院中套院（景园书苑的组合，朝阳公馆的地上与下沉庭院），楼中嵌楼（昌平宅院里的戏楼）。建筑格局因实际建造条件而格外多样。

中国传统建筑通常呈矩形，由若干"间"或"梁架"组成，构成屋顶，每两根立柱间设墙体、门窗或活动隔扇；而花园亭台则四面和顶上完全敞开。这一骨架结构也带来了平面的组合多样化；而高低不平的地势带来了更多的可能性；在处理手法上中式传统建筑崇尚对称布局，十字交叉轴线鲜明。这些传统，在建设条件允许的情况下，健壹营造尽量做到了继承，例如，景园牌楼之后，书苑对称于主干道的两侧；昌平宅院合围前厅的处理手法也是对称的。

然而，实际建造环境有诸多限制和局限，健壹营造在充分考虑整体建筑质量和环境氛围的条件下，在设计和技术上进行了创新和突破。

Interior and Exterior Layout

Among the three JE complexes, many different layouts can be seen. Some are single structures, while others are groups of buildings. Some are redesigns, and some are completely new. There are closed spaces, and open ones. There are independent courtyards, and sets of courtyards within courtyards.

Traditional Chinese architecture is prominently rectangular in layout, with a central axis and symmetrical design. JE has striven to carry on this tradition, such as the reading rooms of Garden's library being found on either side of the main pathway, and the wrapping foyer of Changping Manor is symmetrical as well.

汉白玉栏杆
White marble railings

游廊是庭院之间的有机串联元素
Covered walkways are a natural way to connect courtyards

左图：庭前香炉
The left: A censer outside a doorway

粉墙引路通幽深
Whitewashed walls lead visitors down to the walkway

左图：碧波荡漾的水庭
The left: The tranquil water of a pond

素净别致的空间
Simple and quiet office space

宛如江南园林的游泳池
A swimming pool designed like a southern garden

4

长物赏析篇

Ornamentations

长物之说，主要来自明朝文人文震亨所写《长物志》一书，该书主要讲述了建筑、园林营造与赏析，分室庐、花木、水石、禽鱼、书画、几榻、器具、衣饰、舟车、位置、蔬果、香茗十二个部分，包括衣、食、住、行、用的各个方面。所谓"长物"，原指身外多余的东西，后泛指与生活审美有关的各种建筑空间、园林景致以及器皿用具。

健壹建筑里的长物设置，无论是昌平宅院里的牡丹砖雕，还是朝阳公馆的锦缎窗帘、丝绸靠垫，抑或西山景园的彩绘柱头、绣品陈列、莲花坐佛台具，还是樟木箱改造的沙发座椅，大致可概括为如下环节：

山水景观

中国古典园林重视堆栈假山和开凿水系，以此模拟自然、回归天地。健壹品牌除了盖房子之外，更热衷于筑山、理池，营造山水景观。

假山分土山和石山两类。受建设条件所限，健壹营造是在公馆和景园内适当利用地形，塑造小丘缓坡，以做背景。山石用于砌筑水池驳岸，或单独陈列，或与白砂结合成为枯山水，咫尺之间，隐约可见山峰神采。

水，更是健壹营造灵动精神的源泉。南宋理学大师朱熹说："问渠那得清如许，为有源头活水来。"健壹作品中的水动静结合，景园里既有静谧的大小池塘，亦有涓涓的溪流或珠玉飞溅的水瀑。昌平宅院室内和西山景园中都有溪流萦回，让人有源源不断之感。朝阳公馆的喷泉水池，景园礼堂外的水景瀑布，几处下沉庭院的内墙聚集点处理成水壁的形式，潺潺水流，不但带来清凉洁净，更有丁零乐奏之妙；使得沉静坐落的建筑，萦绕了流动的节奏和生机。

在构成园林的基本元素中，庭院水体为重要元素之一。无论是滋养生命、寓刚于柔、提升活力，还是招引灵气、启迪智慧，水的作用都是不可替代的，它既有观赏价值，也有环保价值，甚至可调控小气候。《黄帝宅经》指出："宅以泉水为血脉。"因此，完美的建筑群落必须配以水体方能画龙点睛。健壹景园的池塘瀑布水景，在天气晴朗阳光普照时，常常会形成彩虹景观，惹人赞叹、欣赏。

北宋郭熙《窠石平远图》
Ke Shi Ping Yuan Tu, Guo Xi, Northern Song Dynasty

The word "changwu" in Chinese, literally "superfluous things", originated from the Ming dynasty book *Treatise on Superfluous Things*, by scholar Wen Zhenheng. The book detailed the design and appreciation of architecture and gardens. Later the term was expanded to include interior spaces, gardens, and other ornamentations that could be enjoyed.

Many of these "superfluous things" can be found in JE's works, such as peony brick carvings, brocade curtains, silk pillows, patterned pillar heads, antique embroideries, lotus Buddha platforms, and even fusion sofas upcycled from camphor wood chests. The objects can be generally divided into the following categories:

Scenery

Chinese gardens tend to use "fake mountains" and man-made ponds and streams, so as to simulate nature. JE is fond of traditional Chinese painting, and loves creating gardens.

"Fake mountains" can consist of either small hills or ornamental rocks. Where space permits, JE utilizes the natural terrain to create small hills, which serve as a background for a garden. Ornamental rocks may be used as decoration for ponds, within a sand garden, or on their own, for a touch of emphasis.

Water is used by JE to create a sense of movement, with a juxtaposition of tranquil ponds and flowing streams or waterfalls. Both the Changping Manor and West Mountain Garden have winding streams, while the Chaoyang Mansion has a fountain and a small waterfall, and several courtyard faces are fashioned into "water walls". These bodies of water offer a refreshing feel to visitors, while also appealing to their aural senses.

Water is an important element of garden architecture. Its vitality and gracefulness are irreplaceable. It can be enjoyed visually, represents nature, and can even help adjust temperature. A perfect complex must include water as a final touch. When sunlight refracts off of the ponds and waterfalls, rainbows often appear, much to the pleasure of those lucky enough to be there at the time.

水上康宁阁，如蓬岛瑶台
Kangning Pavilion, surrounded by water, like a precious island

水，让一切流动起来
Water adds a sense of movement to everything around it

石上积雪，留下大自然的画笔
Snow is nature's way of making a scene more beautiful

The Story of JE Mansion

冬日寒林
A quiet winter day

一方白墙，如纸如框，点缀高树，俨然如画
A whitewashed screen wall accentuated by a single tall tree

绿漪倒映健壹堂，水光潋滟照碧树

Crystal clear waters reflect the lush greenery

◎筑（著）者说：理想中的园林就是立体的山水画。中国写意水墨画，笔法简练，而章法和布局却意味深长；建筑融入诗书画意，意境丰富而繁妙。

⊙The architect/author says: An ideal garden is a three-dimensional scenic painting. Chinese ink paintings feature simplistic brushwork, yet deep impressionistic qualities. Architecture also has a special place in Chinese painting and poetry.

户牖之艺

"户牖"是古代对门窗的雅称。门是建筑的出入口，窗则是采光、通风的洞口。门窗是建筑的重要组成部分。古建筑中的门分为两大类，一种是实心的板门，比较厚重，一般安设在院落大门的位置，门上常有兽头铺首；另一种是隔扇门，比较轻巧，上面附设花格图案，可以透光。窗的种类比较多，包括槛窗、高窗、落地窗等，上面的花格图案更加丰富。此外，还有开在墙上的空门洞和漏窗，在园林中尤为常见。

传统门窗除具体功能外，还是极为重要的审美对象。门窗属于小木作，与大木作构件形成虚实、体积、色彩方面的对比，制作工艺更为精细。其纹饰图案富含文化寓意，并且有等级差异，如清代以菱花的等级最高，民间则以灯笼框和步步锦最为常见，还有冰裂纹、万字、套方、龟背锦等多种图案。门窗还具有类似画框的效果，把门里窗外的亭台、山水、花木固定成一幅立体的图画，这种手法称作"框景"。

健壹营造汇集了古代门窗的各种形式，无论是板门、隔扇门还是各种花窗，应有尽有，而且根据不同的空间主题分别予以设置，比如宴会厅多装设红漆板门或金漆雕龙板门，以示隆重之意；餐厅雅间和客房多用隔扇门，以示随和亲切。窗的形状千姿百态，除了常见的长方形之外，还有正方形、圆形、菱形等。例如宅院每间包房的入口均设石框铺首大门，门框上的雕花图案与其主题一一呼应。如三省堂两侧老人石像，天酒轩雕嫦娥奔月，三友轩雕松竹梅，景福轩雕蝙蝠，合和堂雕如意、锦盒、荷花，不动声色之间蕴含深长意味。

门窗框景也是健壹营造的重要手法。室内外每个空间都以门窗作为过渡，透过月亮门、雕花窗看去，别有洞天，饶有情趣。例如昌平宅院的三层餐饮区楼梯上的巨大木雕花窗，四周透雕松竹梅，窗中可见对面墙上的佛龛和灯座，如瞻西天。

另外，健壹还把许多收藏的木质旧门扇、窗扇用作类似壁挂的装饰物，直接贴挂在内墙之上，辅以专门设计的灯光，虽无出入采光的功能，却好像舞台布景一样，映衬装饰了整体的空间或平面。

隔扇门讲述的光影故事

Light passes through the screen doors with curious effects

Doors and Windows

Doors mainly act as the entrances and exits of spaces, while windows provide illumination and air circulation. In classical Chinese architecture there are two main types of doors: panel door, which is solid and heavy, usually reserved for the main entrances of courtyards, and often feature carvings of animal heads; screen door, which is lighter, often decorated with patterning, and allow light to pass through. Windows are much more widely varied, as are the patterns that adorn them.

Traditional doors and windows, aside from practical functionality, are also an important aesthetic outlet. Serving as a balance with large wooden components, such as pillars and beams, doors and windows allow for a much higher degree of detailing. These designs are full of symbolic meaning, and represent varying ranks of class, such as how Qing officials and the common people used different patterns. Windows and doors can also capture, or "frame", the scenery which can be seen through them.

The JE team uses various kinds of classical-style doors and windows, depending on the themes of different spaces. For example, in banquet halls, red lacquer panel doors are often used for a sense of grandeur, while in private dining rooms folding screen doors give a feeling of relaxation and closeness. Windows include those that are rectangular, square, round and rhomboid in shape.

Scenery framing is an important technique in JE's repertoire. Every interior space includes decorative windows, moon windows, or other unique touches.

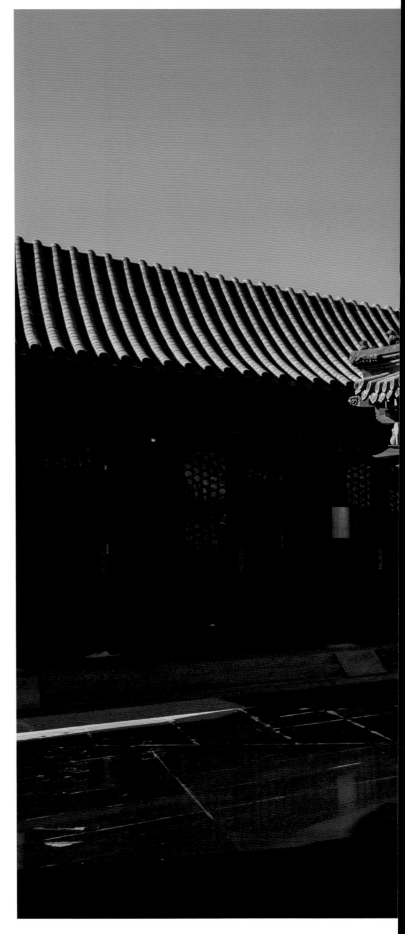

简洁水池倒映传统砖木建筑
The face of a simple pond reflecting a traditional building of bricks and wood

The Story of JE Mansion

隔扇门里藏柜台
A closet held behind a screen door

历史中的旧门扇，海棠花纹漏窗，雕花婚床，从容幽雅

These old doors, ornate windows and carved wooden wedding bed give an elegant and historical feel

右图： 雕花窗扇点缀屏风

The right: A folding screen accentuated with carved patterning

木雕月亮门，斜纹花格窗，翠竹灯影，从容幽雅
A wooden moon door, with a diagonal screen window, through which a bamboo thicket can be seen

◎筑（著）者说：推敲一扇门，开启一片窗；空间有衔接、出入、采光、壁挂和装饰。我本人对旧式门窗情有独钟，先后收集了六千多扇，它们被设计安排到不同的空间里、局部上，希望人们在我们的建筑中，可以小小地博览一番，领略"传统户牖之艺、之用、之美"。

⊙The architect / author says: Doors and windows connect rooms, provide light, and are pleasing to look upon. I have a personal affinity for antique doors and windows, and have collected over 6000 of them. They are used throughout the structures we built, and I hope that visitors can get a sense of the art, usefulness and beauty of traditional Chinese doors and windows.

左图：镶铁装饰，古朴凝重
The left: Inlaid iron makes for a rustic and solemn appearance

家具陈设

室内的空间效果主要依靠装修陈设来营造，对此健壹营造也是巧思设计，深下功夫。

以朝阳公馆客房为例，其中的豪华套间常被客人订为婚房，分为里外三间，横长的外厅居中设有沙发，中室设大床、柜子和座椅，最里面是临床的洗浴间，环环相扣。还有一间利用较高的室内空间高度，借鉴传统建筑"仙楼"的手法，搭建了一个夹层阁楼，以设书房，铺设竹席、棋桌，飘然俗境之外。

为与每个房间的主题相呼应，屋顶天花精心设计，大多借鉴古代井口天花的式样；有的则采用相对简洁的现代手法，与各种吊灯、顶灯结合。

室内隔断包括板壁、花罩、博古架和碧纱橱四种形式，可以灵活分隔内部空间：

板壁，南方又称太师壁，常置于堂屋正中，上可雕大幅图案，也可以悬挂卷轴的书法作品。

花罩，类似于门框的效果，样式很多，健壹营造常用的是圆光罩、落地罩和飞罩。

博古架，用木板分成很多大小不等的格子，用来陈列古玩和工艺品，也可用作书架，给室内增添清雅的气氛。

碧纱橱，与隔扇门的形式相似，用在内部的隔断上，一般除中央两扇可开启外，其余固定，其内芯部分常常两面夹纱，上绘图案或书法。

《红楼梦》中林黛玉所住的潇湘馆，"里面都是合着地步打就的床几椅案"，意思是说其家具陈设与铺地图案一一对应，极显匠心，可令今人叹为观止。健壹营造同样也做到了这一点，以昌平套房为例，其地毯、铺地就与桌案、沙发、书架完全对应，浑然一体。

大量的旧式床、桌、椅、柜陈列在大堂、餐厅和客房中，很多家具还带着磨损的岁月痕迹，别具沧桑美感，带来悠远的历史味道。强调使用功能和舒适性的西式家具与中式家具一起搭配摆放，各安其位，毫无违和感。

墙上的装饰除了旧式门窗之外，还有大量名家字画、京绣、杨柳青民间年画。既是主人精心的收藏，也是民间艺术的拾珍。

书柜中放置的大多是线装书和精装套书，博古架、桌案上陈列着许多文玩器皿，都让人爱不释手。雅集鉴赏把玩，书房挥笔泼墨。

还要特别提一提健壹营造中的灯具，无论吊灯还是台灯，大多为品牌特制，造型极为多元，或以佛像为灯座，或呈莲瓣形，或如大红灯笼高挂，或如纱罩座灯落地，往往都蕴含着特殊的艺术主题和特别的寓意。

Furniture

The creation of interior spaces mainly depends on the selection and placement of furniture. JE spends much time and effort on such designs.

Reflecting the themes of each room, their respective ceilings are carefully designed accordingly, with layered lattice patterns for most, while others use simplistic modern styles, combined with hanging lights.

Interior partitions include panel walls, flower covers, display shelves and gauze cabinets, which separate sections of rooms will adding to their overall atmosphere.

The home featured in the classical novel (*The Story of the Stone*) has been described as having furniture that corresponds with the patterns on the floors. JE has striven to achieve this as well.

Much of the antique furniture in the halls, dining rooms and bedrooms bears the wear and tear of many years, for a time-worn feel. This is further combined with modern furniture intended for comfort and practicality.

On the walls are many works of calligraphy, painting, embroidery and woodblock printing. These are all from JE's collections, and are highlights of Chinese folk art.

Throughout the rooms are many shelves which display a wide variety of antiques, providing visitors with a unique and memorable experience.

坐床靠椅倚墙而设
Sofa and chairs are placed against the walls

The Story of JE Mansion

华堂美厦，陈设典雅
A spacious room with elegant furnishings

右图：豪华气派从地而起
The right: The grand feeling of this room begins with the carpet

灯具造型
Ornate lamps

左图：充满个性的茶几
The left: A one-of-a-kind tea table

The Story of JE Mansion

温馨舒适的细节陈设
Meticulous details in furniture selection

京绣点缀墙面
The embroidery adorning the walls

灯具造型
More ornate lamps

◎筑（著）者说：灯光与照明、家具陈设可灵活多变地定义建筑空间的功能，并点缀烘托整体风格；强调重点、点亮渲染或者映衬背景，使用手法是非常多元化的。

⊙ The architect/author says: Furniture arrangement is a versatile way to determine the functionality and theme of interior spaces. It can be used either as a highlight or a background, and we have a wide variety of techniques to choose from.

The Story of JE Mansion

绿植花木

植物是中国古典园林中另一种不可缺少的景观要素。受原有条件限制，昌平宅院以封闭室内为主，较少花木布置，朝阳公馆和西山景园都拥有丰富的植物景观，四季可赏，充分展现了大自然融合映衬建筑的魅力。

健壹营造所用的植物品种主要包括松、柳、槐、石榴、柿子等乔木与海棠等灌木，以及竹子、草坪等；盆植花卉相对有限，主要用于室内点缀装饰。

其中松树以油松为主，枝干挺拔，树冠如盖，浓郁常绿。例如健壹公馆西南下沉庭院中央种植的一株松树，如龙蟠凤舞，翠盖掩映，坚韧、常青、虬劲。

柳树多用北方常见的绦柳，形态近似于南方的垂柳，一般种于水池岸边，其小枝柔软倒垂，树叶狭长。健壹景园池边随处可见杨柳依依的绿丝绦。

槐树树干虬劲，枝叶繁茂。龙爪槐更是造型独特，树冠浓密，为景园书苑室外的主要树品。

竹素有君子之喻，是一种多年生禾草类植物，生长迅速，其茎干为木质，坚硬、多节、中空，一般外观通体呈碧绿色。

在中国古典文化中，绿植花木不仅是情感载体，还是文化符号和吉祥如意的象征。这一特点体现在健壹园林之中，就是寓意深长的选择和置放。例如，西山景园华府区，紧邻中式婚礼堂，专门移栽了一株石榴，寓意多子富贵的美好祝福；翠竹成排，虽为室内，却宛然露天庭院；柿树果实累累，象征事事如意。

健壹营造中的盆栽花卉，主要依照北方四季物候特征，按建筑空间功能区域，布置有兰花、水仙、富贵竹、玫瑰、月季、菊花、文竹等，或清新可人，或芬芳馥郁，雅致情趣无处不在。

中国本土园林一般没有大片草坪，健壹公馆和景园的草坪均为西式园林手法，采用进口草种，能够在北方天气中保持较长时间的青翠。

在作为庭园景观重要装饰之外，绿植花木还起着调节空气等的特殊作用，例如消减现代室内各种辐射、静电，通过光合作用释放氧气，为建筑提供新鲜空气。

这些竹木花卉虽属配角，但具有重要意义，春萌夏繁，秋收冬藏，赋予建筑、环境别样的色彩体验和微妙的季节变化；它们与亭台楼阁、山石溪流相映成趣，共同把建筑这一立体画布渲染得更加绚丽。

Trees and Flowers

Plants are an essential element of classical gardens. JE's works feature ample greenery, which changes depending on the season, and greatly adds to the juxtaposition of natural scenery and man-made structures.

JE uses trees such as pine, willow, scholar tree, pomegranate and persimmon, as well as shrubberies, bamboo and grass. Potted plants are used less widely, and mainly as accents to interior spaces.

In Chinese traditional culture, trees and flowers are not only used for aesthetic reasons, they also embody auspiciousness and other cultural symbolisms. For example, next to the wedding hall of the West Mountain Garden, a pomegranate tree was chosen; pomegranates are known for their large amount of seeds, and in Chinese "many seeds" is homophonous with "many children", hence its cultural significance as a marriage blessing.

The potted flowers used are mostly northern varieties, and include orchids, daffodils, roses and Chinese roses, chrysanthemums, and so on, adding color to interior spaces.

Most Chinese gardens tend to have no broad, open lawns; but JE Mansion includes a particularly large one, consisting of imported grass, which stays green throughout most of the year with the northern climate.

Aside from an integral element of gardens, trees and flowers can also adjust air quality, by producing oxygen through photosynthesis, and reducing the radiation of static electricity.

These plants and trees may not be the centerpieces of a complex, but carry important significance, as their changes through the seasons alter the color and appearance of the environment. They form a harmonious balance with structures like pavilions, as well as decorative rocks and flowing streams, creating an experience like a scenic painting that one can walk through.

花台
A raised flower bed

The Story of JE Mansion

191

绿草如茵
Lush green grass

The Story of JE Mansion

193

绿草碧水相交映
Greenery reflected in a tranquil pond

◎筑（著）者说：古人云，长物，不是矫情之作，而是性情之作。以前的文人雅士希求在一个自己参与设计和建造的宜居环境中，过一种雅致的读书生活，非为了修身治国平天下，而是一种颐养。

"物物而不物于物"的平衡之法，在现代的今天，讲究生活质量与内涵的人，在紧张与休憩、严谨与随意、入世与出世的现实光阴里，追求的和古人是近似的，即通过对物的调遣、安置和鉴赏，而达到心灵的平静与和谐。"非有真韵、真才与真情以胜之"，长物不在路数和消费的物质表象，而在于韵、才、情的合一体现。

⊙The architect / author says: The ancient Chinese sought to create a comfortable space where they could live a scholarly life, not in preparation to contribute toward the nation, but purely for self satisfaction. In our modern society, those who pursue a life of quality and meaning, whether it be during work or rest, formal or casual, have a frame of mind similar to the people of ancient China. This is reflected in the "superfluous things" we choose to place in our living spaces, not in how much we spend on them, but how we enjoy them.

The Story of JE Mansion

5

传承创新意蕴篇

Heritage and Innovation

古人云："治大国若烹小鲜。"对于健壹营造而言，则是筑者所谓"营造建筑仿佛调配五味"，其营造的系列作品在秉承传统、发扬经典的同时，又创新突破，充分考虑现代人们生活的环境和需求，运用现代技术和手段，与时俱进，精心呈现了符合现代需求的建筑作品，营造出典雅优美的环境，可谓融汇古今、沟通南北、合璧中外。

融汇古今

近代以来，中国的木构建筑日渐式微，今天我们的现代建筑基本上已经是西方建筑的势力范围，传统的木结构方式甚至被看作落后的象征。但实际上，中国古代木构建筑在模数化设计、预制装配式的施工方式以及造型、空间方面，恰恰与现代建筑的不少理念有契合之处，而且当前国外一些新建筑对木结构又有新的发展，可见木这种材料历久弥新的旺盛生命力。健壹营造继承中式建筑的木之传统，大量使用木结构，取得了良好的建造效果和别致的艺术美感。

传统的建筑造型使得健壹营造具有强烈的识别性：厅堂、重檐方阁、水心榭、亭子、垂花门、戏台，一梁一柱，一砖一瓦，各有各的妙处，令人充分领略传统的营造之美。从浙江搬迁而来的老房子不仅保存了文物，而且带来了醇厚的古典氛围。例如，健壹景园中书苑东院的大堂，完全由一座浙江金华地区的清代民居老宅搬迁而来，其木构架采用典型的穿斗式，纤细灵巧，其间穿插牛腿、月梁、木栏杆，雕饰精美。中央的天井小院安装了玻璃顶棚，成为大堂中庭，四面环列木柱，南为隔扇影壁，左右各设木楼梯通向二楼书房，北面正厅居中立屏风，前设坐床和沙发，空间规整堂皇。前廊和正厅的天花板三处上凹，形成类似藻井的效果。檐下悬挂的四面旧匾额，更显古朴典雅之风。

景园客房区大堂同样是一座古民居，门厅、左右廊与正厅围绕一个狭小的天井，格局紧凑。正厅中央设有旧式火炉，四周摆放沙发，两边分设散座、小几和书柜，室内檐下各处悬挂着"德荣指使""稀龄衍庆""眉案双清"等古旧匾额。而酒廊吧台的设计，采用中式窗棱镂空深色木饰，后辅以现代亮色暖光照明，既现代时尚，又含几分古意。

大量的旧式门窗安装在外墙与内壁，几乎每个房间中都陈设有古典老式的柜、床、桌、椅等家具，匾额、楹联、书画、唐卡、京绣、年画无一不浸透了岁月的沉淀，凝固着历史的文化气息。

但是，学习古人绝不意味着消极保守和故步自封。我们今天的生活方式已经发生了天翻地覆的改变，建筑本身也必须满足现代人的多元需求。健壹营造在传承历史的同时，融入了大量的现代元素。旧楼的混凝土结构得到保留，所有的建筑功能设施，照明、排水等，采用先进合理的现代技术与设施；同时，无论整体造型、内外空间还是各种细节上，注重古典传统的艺术形式和装饰，体现了古今融合的特点。

于是，我们看到古色古香的方亭下围合着透明的玻璃栏杆，现代酒柜与古代长条桌案共处一堂，新式沙发与老式座椅相互依偎，收纳空间掩映在雕花门窗之后，简洁的方盒子中隐藏着整座四合院，平屋顶上耸起一片翘角飞檐。

原来，古与今、新与旧，并非格格不入，而是如此相得益彰。

A thinker in ancient China said, "Governing a state is like cooking." For JE, "Architecture is like cooking" is the case. JE's works, while carrying on the legacy of tradition, also seek to innovate, by using contemporary technology and techniques to cater to the needs of modern society. Each of their projects fuses past with present, north with south, and East with West.

Past and Present

In modern times, wooden architecture has been all but absent in China, with virtually all new buildings being constructed using Western methods, and wood even being considered by many to be "a thing of the past". But actually the designs, capabilities and principles associated with wood conform quite well to modern standards, and many new buildings around the world have used wood in new and inventive ways. JE uses large amounts of wood, to carry on the heritage of traditional Chinese architecture, while achieving unique artistic effect.

Traditional architectural designs have given JE's works an extremely distinctive appearance.

The main hall of JE Garden's guest quarters is a traditional home, featuring a terse design in which the main chambers and corridors wrap around a central skylight. In the foyer are an old-fashioned fireplace, a collection of sofas, tea tables and bookcases, and wooden calligraphy tablets hang on the walls. There is also a bar, made out of dark-colored wood in a Chinese style, illuminated with modern lights.

The interior and exterior walls feature large numbers of old-fashioned windows, and in each room are collections of antiques, such as antique furniture, couplet scrolls, calligraphy, paintings, embroidery and wood block printings.

However, learning from tradition by no means signifies extreme conservatism and refusal to progress. Our lifestyle today has undergone significant changes, and architecture must cater to our modern needs. JE, while carrying on tradition, also adds many modern elements. All functional facilities, lights and piping use the latest technology, while old-style concrete frames are still used, along with traditional artistic decor in all the rooms.

Thus we can see things like an old-fashioned pavilion surrounded by a glass railing, a modern bar next to antique tables, modern sofas interspersed with upcycled old chairs, and an indoor quadrangle courtyard with latticework windows.

The purpose of fusing the past with the present is not for them to stand out, but complement one another.

右图：虬枝苍劲
The curled branches of an old tree

健壹营造把传统音符融入现代乐章
Adding a touch of modern living to traditional aesthetics

健壹营造志

The Story of JE Mansion

最时尚的红酒柜，采用最传统的木工手法
A fashionable wine cabinet, made using the most traditional handcrafting methods

现代沙发与传统桌案共存一室
Modern lounge ware and traditional tables are placed together in the same room

舞台前厅，别致又隆重
The entrance to the theatre, distinctly elegant

古雅的小亭与简洁的墙面彼此映衬
A rustic little pavilion and plain, unadorned brick walls complement each other

现代器具与传统装饰背景
Modern facilities with traditional decration

沟通南北

中国版图辽阔，各地风土人情千差万别。从地理角度而言，地方和人都有南北之分。北方文明由黄河流域孕育而成，多显雄浑大气、稳重端庄；南方文明以长江流域为代表，以清秀婉约、细致精巧为特征。从建筑的角度来看，南方的民居、园林大多空间曲折，色彩素雅，常见粉墙黛瓦，屋檐上翘明显，富有灵秀精致的气质。北方的宫殿、府宅、园林大多空间宽敞直爽，色彩偏向厚重，庭院相对宽阔，具有端庄大度的气质。

健壹营造的三个作品都坐落在北京，建筑样式主要采用北方的官式做法，屋檐曲线平缓，庭院和室内空间格局以对称为主。外墙主要使用灰砖，木构件局部点缀浓艳彩画，一方面避免沙尘、雾霾的污染，另一方面相对弥补漫长冬季色彩单调的肃杀感，同样也反映了北方地区的直爽性情与鉴赏习惯。

中国的南北文化既有区别，也有相互影响的一面。二者在历史发展进程中不断碰撞和交融。由于大运河的开凿，历代定都于北方后均大量吸收来自南方的物资、人才及文化因子。尤其在清代，宫廷经常征召苏州、扬州等地的能工巧匠进京，为皇家和上层豪门建造宫殿和府邸，康熙、乾隆两位皇帝还热衷于在皇家园林中模仿江南名园，使得北京的建筑中渗透了许多南方元素。

健壹营造也继承了这一传统。其亭台楼阁以北京固有的建筑形式和抬梁式构架为主，又吸纳了江浙和安徽地区的很多古建手法，比如部分山墙借鉴徽派民居高大的"马头墙"造型，部分室内墙面采用江南特有的粉墙漏窗，木构件大多采用原木色，局部雕饰具有强烈的南方韵味，很多旧式门窗、家具、匾额、楹联都来自南方，还直接将几座浙江古民居镶嵌在整体建筑之中。

园林景观方面，健壹营造更多学习了江南园林曲折多变的空间路径，小桥流水，湖心阁楼，一块瘦、皱、透、漏的太湖石，两三竿挺直幽响的翠竹，一曲蜿蜒流淌的小溪，几池蓬莲相映的荷塘，与天井、小院、游廊相互穿插，淡雅江南风情，娓娓道来，挥之不去。

按照中国古代的相面之说，"南人北相"或"北人南相"都被认为是有福之相，原因可能是这两类长相都融合了南北之长。同样道理，如果南方建筑在婉约中融入一丝大气，北方建筑在端庄之中渗透一点灵气，即可取得良好效果。健壹营造是典型的"北建""北园"，而又营造得不乏"南相"，故而难能可贵。

北方建筑的厚重感
The bold appearance of northern architecture

North and South

China covers a broad expanse of terrain, with widely varying cultures and customs. Geographically, China can be divided into the two major regions of north and south. The civilization of the north sprouted from the banks of the Yellow River, the people there enjoying majestic and direct styles. The south is represented by the people along the many waterways in the basin of the Yangtze River, who prefer a more refined and demure lifestyle. In terms of architecture, southern homes and gardens use many winding spaces, with a simple yet elegant color scheme, and fine attention to detail. The palace halls and manors of the north, on the other hand, emphasize spaciousness, grandeur and vivid colors.

All three of JE's projects to date are located in Beijing, and constructed according to mainly northern official-style designs, with gently sloping roofs and symmetrical layouts for interior and exterior spaces. The walls consist chiefly of traditional gray bricks, and in some places wooden components are colored in order to stand out. Overall, the structures give a sense of the northern Chinese persona of boldness.

Northern and southern Chinese cultures bear both differences and commonalities, and through the centuries the two have constantly been intertwined and fused. And with the implementation of the Beijing-Hangzhou Grand Canal, the capitals of each dynasty, most located in the north, were exposed to the resources, people and culture of the south, thus many elements of south architecture can be seen in Beijing.

The pavilions and other small structures in JE's works are mainly designed based on the Beijing-style raised platform framework, combined with traditional techniques from the south. For example, one can find the protruding "horse head walls" of Anhui, whitewashed walls characteristic of the region south of the Yangtze River, and many of the wood carving styles are from southern China, as are a large number of the antique doors, windows, furnishings and wooden tablets.

As for garden design, JE mainly drew upon the winding garden paths, trickling streams, decorative rocks and bamboo thickets of the south for inspiration. Ponds and small lakes can also be enjoyed, many of which have pavilions accessible by stone bridges. Covered walkways, skylights and small courtyards are also all very southern touches.

According to ancient Chinese divination, being a "southerner with the look of a northerner" or vice-versa was considered to be lucky, as one's persona combined the best of both worlds. Likewise, the architectural characteristics of north and south can also be fused together for stunning effect. JE strives to achieve designs that are fundamentally northern, accentuated with qualities of the south.

右图：书苑中镶嵌的江南古宅
The right: A distinctly southern-style chamber

青山绿水,曲折通幽
A quiet path winds among the scenery

The Story of JE Mansion

The Story of JE Mansion

合璧中西

中国自古以"上邦天朝"自居，但历史上也不断受到外来文明的影响，早期的佛教传入，明清时期的西学东渐，都给华夏文明注入了新鲜的元素。欧美文化的传播带来更多的西式建筑与园林风格。清乾隆年间，曾在圆明园中模仿欧洲巴洛克风格兴建过一组西洋楼，江浙、广东等地的富商住宅、民居也经常使用西式装饰、喷泉、玻璃镜，可谓中外合璧的先驱。

关照现代中国人的生活方式，结合古典传统，营造出符合现代生活方式，又兼具中国古典韵味的建筑环境，健壹在秉承中国传统的同时，也甄选地吸收了外国建筑的优良元素，融合采纳了异域园林的某些处理方式。

健壹景园的中式礼堂正是中西风格融合的典范。其九开间前后廊的木构架属于典型的传统抬梁体系，但不是像古代建筑那样强调以进深方向为纵向，改以面阔方向为纵向，很自然地把室内空间划分为中厅与两旁侧廊三个部分，外墙设置大面积落地窗，直接借鉴了欧洲巴西利卡式大教堂的空间形式，特别宜于举办现代婚礼，同时又通过花轿、楹联等一系列元素强化了中国本土文化的特色韵味。

在健壹建筑的许多细节上都可以发现西式风格的影子。比如房间家具的搭配，现代的床具床垫会覆盖传统的刺绣床单，既舒适又保持老派的风范；欧式沙发、酒柜经常与中式坐床、圈椅、几案摆放在一起；天花板的装修有时会借鉴欧美的样式，而灯具的选择设计会是中式底座配以西式纱罩，混搭出别致又舒适的风格质感，以求更好地配合建筑功能的使用，并与空间的主题相契合。

在园林景观方面，除了纯粹的中国古典园林手法外，健壹营造主要吸收了欧洲园林开阔的草坪以及喷泉水池，并在局部种植修剪整齐的灌木，设立雕塑。此外，健壹还在内外庭院、屋顶平台、连廊过道等处，多次采用日本"枯山水"园林的手法，即一种抽象化的园林方式，以砂代水，以石代山，以极其简单的素材代表自然景物，蕴含着浓厚的禅宗意味，与中国传统意境不谋而合。

左图：高大的马头墙是徽派民居的特色
The left: This style of raised horse-head wall is a staple of southern Hui residential architecture

内庭枯山水
An indoor rock garden

East and West

Throughout history China has been constantly influenced by outside civilizations, such as Buddhism in the Han dynasty, then Occidental culture later in the Ming and Qing, including Western-style architecture and gardens. Yuanming Yuan, built in the Qianlong period of Qing dynasty, featured several Baroque-style buildings, while in regions such as Jiangsu, Zhejiang and Guangdong fusion homes bore Western ornamentation, fountains and mirrors.

JE, while carrying on Chinese tradition, also selectively absorbs and implements some elements and methods of other styles of architecture and gardens.

The JE Garden's Chinese-style wedding hall is representative of Chinese and Western fusion. The rafters of its corridors are made with traditional raised beams, but unlike the old fashioned style, the beams taper outward rather than inward, naturally dividing the interior space into three sections. Most of the windows are of the Basilica-style floor-to-ceiling variety, perfect for weddings, while at the same time elements such as a bridal sedan chair and couplet scrolls strengthen the sense of traditional Chinese culture. Furnishings also involve a great degree of fusion, such as modern mattresses covered with traditional embroidered sheets; European sofas and bars juxtaposed with Chinese couches, round stools and tea tables; and Chinese ceiling lights with Western-style gauze covers, creating an atmosphere that is unique yet functional, with a consistent and unified theme.

As for gardens, JE's works include distinctively Chinese gardening techniques, as well as European lawns, fountains, neatly trimmed shrubs and sculptures. In addition, there are also interior courtyards, roof terraces, connected corridors, and numerous Japanese-style rock gardens, the latter representing a mountain island in the sea, a type of symbolism which is very similar to Chinese artistic conception.

右图：日式枯山水布局
The right: A Japanese rock garden

古代样式的铺面成为最新潮的吧台
An old-fashioned shop front has now become a fashionable bar

福字墨宝下的钢琴
Chinese calligraphy on the wall next to a piano

玻璃穹顶下的中式木作桌椅
Underneath a glass dome, a Chinese-style wooden table and chair set

The Story of JE Mansion

◎筑(著)者说:房屋建筑,终究是为人来使用居住的;因此,人在建筑里的体验感受和舒适程度,是我所关注和要关照到的。中学为体,西学为用;无论哪一种传统和风格,一定要带给人舒适和美感,才会富有生命力而得到传承和继续。另外,在采用不同风格和传统时,设计者应具备大局整体的协调观,不是好东西堆在一起就是更好,取舍、搭配,也是设计者修养和创意的体现。我们的健壹系列,也是在不断探索和实践中,逐步建立信心,形成风格的。

⊙The architect / author says: Residential buildings are designed for people to live in, thus my focuses for such spaces are people's experiences and comfort levels. The overall theme is traditional Chinese styles, with Western execution. Regardless of the tradition used, the space must give one a sense of comfort and beauty, as only this way will such a style be carried on. Also, when using different styles together, the designer must be able to look at the whole picture, not just any nice items will be better together, and knowing what to pair and what not to pair is a reflection of the designer's skill. In our projects, we constantly explore and experiment with different combinations.

左图:西式水池与中式墙体
The left: A Western moat and Chinese-style wall

哲匠精神

所谓"哲匠",是指有思想、有文化而又技艺精湛的匠人,也是对古代建筑师的最好褒奖。健壹营造的系列建筑,深刻而又生动地体现了古代的哲匠精神;在塑造空间、磨炼细节和追求文化意境等方面极下功夫,并且赋予了这些建筑内容以新鲜的时代内涵。

创意空间

老子《道德经》中说过,建造房屋,本质不在于屋顶、墙壁、梁柱这些实体的部分,而是用实体围合而成的虚空部分。从这个意义上说,实体是建筑的外壳,空间才是建筑的灵魂。健壹营造对塑造空间极为重视,展现出丰富的创意。

健壹作品的整体格局和主要空间,大多拥有明确的中轴线,无论是昌平宅院中的室内四合院、朝阳公馆的楼堂布局,还是西山景园中的各个组成部分,都讲究对称均衡,端庄有序,明亮大方。同时其中的餐厅、客房、庭院的朝向、形状、尺度各不相同,彼此排列组合,幽明互补。山水、植物、装饰小品穿插其间,加上日光、灯影的映射,增添了许多空间的艺术化效果。

朝阳公馆大堂的空间处理是健壹最注重的创新示例。原本大堂空间只有一层高度,且改造完成后始终不太理想,后来接受了一位海外建筑师的建议,在二楼中央开辟了一个八角形的天井,屋顶上增加了一个采光的方亭,变成通高三层、上下贯穿的形式,于是"感觉一下子就亮了"。这一变化虽然牺牲了部分平面经营空间,但成为艺术空间的神来之笔,所有客人都对此处印象深刻,赞不绝口。

又如昌平宅院的戏台前,既有二层高的散台大厅,又在三面设置半开敞的格子间和二楼包房;西北角上还布置了两座方亭和一湾溪流,空间有开有合,有聚有散,高下参差,丰富多姿。这不是传统的戏楼建筑布局方式,却在现实探索中既满足了多重功能,又取得了良好的效果。

健壹建造中改建部分的餐聚包间和客房,大多属于相对封闭的小空间,形状差异很大,其设计并没有单调统一处理,而是充分利用原有的平面进行分隔,异形的角落往往设置为备餐间、卫生间,主体部分纷纷呈现出

各具个性的空间特色，如回字形、L形、C形等，各有妙趣，极少雷同。如昌平宅院三层中央位置有四间客房无法对外采光，特意在彼此之间设置四个微型庭院，点缀山石、翠竹，使得房间转化为情趣盎然的景观房；一间卧房在床后巧设一装饰性背龛，几株芭蕉掩映其间，立刻增添了许多清雅气息。

建筑物中的过厅、走廊和楼梯等过渡性空间，往往是设计者难得泼墨施展的地方，而健壹营造却对这些看似配角的地方毫不忽略，巧妙地在其中或点缀一方佛龛、两株花竹、几扇门窗，或稍加隔断，悬挂匾额、楹联，使得咫尺之地，也变成一道令人驻足欣赏的景观。例如朝阳公馆走廊内，铺设了专门的寿字地毯；在二层客房区的过厅，特意安装了玻璃采光屋顶，一缕阳光洒在壁间《无量寿佛图》上，成为一个独立的小型佛堂，客人过此，尘心往往为之一洗。

室外庭院空间同样富有特色。如昌平宅院利用屋顶平台营造出小庭院若干，氛围静谧安详。朝阳公馆的下沉庭院呈台地式，隔墙月亮门上有"品竹"二字匾额，院内沿墙种竹，幽篁如玉；东院中建有一座方亭，相邻小院方池之上砌筑方形平台，四角立石雕望柱，池底铺设卵石。西院则在水池中央种了一株松树。首层北侧房间均朝向大草坪开门设窗，视野开阔；南侧房间则与庭院相同，独享宁静。正所谓"奥如旷如"，各有妙处。

The Spirit of Craftsmen

In ancient China the highest title that could be granted an architect was "Master Craftsman". The spirit of such craftsmanship is ubiquitous throughout all of JE's works. The team took great efforts in all aspects, such as space design, detailing and cultural atmosphere, to granting the spaces a refreshing feeling of timelessness.

Laozi, in his *Tao Te Ching*, said, A home consists not of the physical elements such as its roof, walls, pillars and beams, but of the immaterial space created within these parts. In other words, the building itself is merely the shell of the home, the space inside is its true "soul". The JE team greatly values the creation and innovation of spaces.

The overall layout and main spaces of JE's complexes mostly center on a clearly defined central axis, emphasizing symmetry, balance and order. In addition, the rooms and courtyards are designed in a wide range of directions, shapes and sizes, complementing one another and forming a whole. The artistic conception of these spaces is further realized through plants, decorations, and both natural and artificial lighting.

Chaoyang Mansion's main hall is one of JE's most representative innovative spaces. The hall was originally one floor, but later, upon suggestion from an architect from abroad, JE made an open octagonal space in the center of the floor, and a skylight above, so that all of a sudden "everything became much brighter". Although some floor space was

sacrificed, this change in the hall's design added an effect that has been praised by countless visitors.

Another example is the Changping Manor opera theater, which features a scattered foyer, with half-open rooms on each side, and private rooms above. In the corner there are also two small pavilions and a stream, so that the foyer as a whole contains a varied environment juxtaposing open and closed, collected and scattered, and high and low. These features are not found in traditional opera houses, instead having been created out of practical need, satisfying both functionality and aesthetics.

Transitory spaces, such as connecting halls, corridors and stairwells, are often places that prove challenging for designers. But JE did not ignore these often neglected yet important building components, infusing them with subtle touches such as a Buddhist altar, a couple of bamboo plants, windows or doors, a hanging wooden tablet, a pair of scrolls, and so on, fashioning these small spaces into sights to be enjoyed as one passes by. Exterior spaces are equally distinctive, such as the numerous balcony courtyards of the Changping Manor. The Chaoyang Mansion features a lowered courtyard, with a moon door, wooden calligraphy tablets, and bamboo along the perimeter. The east, west and north courtyards all possess different designs and views, each offering a unique experience.

婚礼堂的"十全十美"
The captivating design of the wedding hall

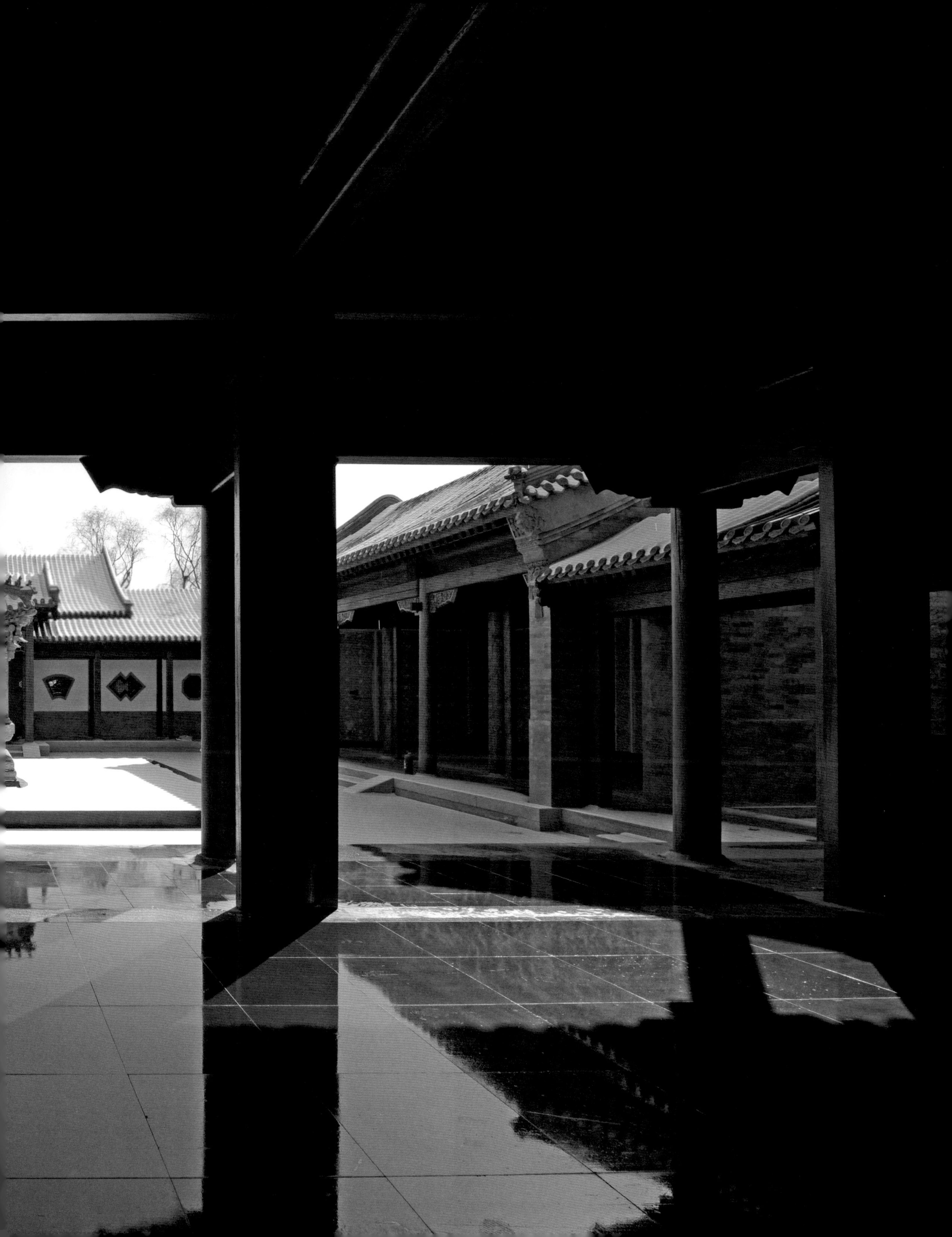

下沉庭院带来明亮的窗户背景
The lowered courtyard design grants a bright view of the exterior

右图：咫尺庭院，观天俯地
The right: The terse layout of this courtyard offers a view of both above and below

佛光普照
Light plays through the screen windows

戏台对面闹中取静的包房
Across from the opera stage, a private room offers quiet and privacy

精工细作

现代建筑大师密斯·凡·德·罗曾经说过:"建筑的魔鬼在于细节。"建筑毕竟不是纸上的书画或弦上的旋律,而是实在立体的客观存在,由台基、柱、梁、墙、屋顶组合而成,再好的设计和构思,也需要精心的施工才能最终得以实现。

健壹营造的系列建筑,展现出当今十分罕见的工匠作风,精工细作,务实踏实,又追求唯美,每个细节都力求极致完美。本人坚持亲自在工地上指挥施工建设,正是为了追求和达到尽善尽美。有一次参观古建筑,大家感慨古代砖墙砌得严丝合缝,而现在已很难看到如此的手艺时,我却自豪地说:"我们的房子就能砌成这样。"这句话的背后是对自己建筑作品和建设团队之精工细作的信心。

古代的木构件加工、砌砖、码石、布瓦都有一套完整的做法,充分展现了传统的工艺之美。以"干摆"式砖墙为例,砌筑时要选择大号的停泥砖,把砖的五个面都做砍削打磨,形成梯形断面,每铺好一层后,浇灌一种用白灰和黄土调制的桃花浆,然后再铺上一层,每铺五层就需要晾一段时间,等灰浆基本凝固以后再接着铺砌。砌完后还得对墙的表面进行反复打磨,保证平整,最后用水冲洗,显得特别干净利落。朝阳公馆和西山景园的墙体外观、楼体立面,都是这样精工细作的典型样例。

健壹营造的木构件坚持采用榫卯方式连接。"榫"指的是构件上附带的凸起部分,"卯"指的是凹入的开口部分,两个需要组合的构件一个带榫,一个带卯,彼此一迎合,就严密地卡在一起了。这种构造方式特别合理,留有若干伸缩的余地,属于柔性的连接,遇到地震、大风时可以减少危害。这使得健壹作品有着区别于其他现当代建筑作品的独特而别致的古典韵味。

另外,健壹营造中还拥有大量的木雕、砖雕、石雕,其中有些源自老房屋的部件,更多是现代匠师的杰作。木雕以浮雕为主,有时候也会出现富有立体感的圆雕和镂空的透雕。透雕的效果类似剪纸,大多用于室外的吊挂楣子和室内的花罩。室门外墙壁上往往凸显大幅砖雕图案和书法,银钩铁划,气韵流转。石雕主要见于门墩、门框、栏杆以及成对摆放的麒麟、狮子,都镌刻得十分生动。即便是属于现代风格的结构、装修、陈设,健壹营造同样做得一丝不苟。

天井打通上下,点亮整个大厅
A skylight brings ample illumination to the foyer

The Story of JE Mansion

The Details

Modern architect Ludwig Mies van der Rohe said: "In architecture, the devil is in the detail." Architecture is three-dimensional and objective, composed of foundations, pillars, beams, walls and roofs, and even the best design requires meticulous attention to detail in order to feel complete.

JE's series of projects present an overarching architectural style quite rare today, with perfect sought in every detail. I insist on being at the building site throughout construction to guide the workers, so as to ensure every aspect is created as intended. One time while visiting an ancient structure, everyone was amazed by how tightly the bricks were laid together, a phenomenon rarely seen in modern architecture, to which I responded, "The bricks of our buildings are like that too." This statement served as a testament to my confidence in my team.

There is a complete methodology for ancient Chinese wood work, brick laying, stone placing and tile setting. Bricks used for walls had to be selected, cut, sanded and laid according to certain rules. These methods were used for all the brick walls of the Chaoyang Mansion and West Mountain Garden, as homage to these meticulous traditions.

The wooden components of JE's structures are connected by traditional "mortise and tenon" joints. These convex and concave pieces are tightly locked together, in a sturdy yet pliable crisscross formation, granting the structure a characteristically classical appearance.

Throughout JE's works one can also find large numbers of wood, brick and stone carvings, among which many are antiques collected from across the country, while others are works by modern craftsmen. For wood carvings relief is the most common style found, often featuring stunning raised circular engraving and hollowed out parts.

一丝不苟的装修
Relentlessly meticulous ornamentation

左图：磨砖对缝，描红画彩
The left: The bricks are carved to perfect precision, then colored by hand

The Story of JE Mansion

健壹营造志

◎筑（著）者说：精工细作，代表的是敬业精神，传承的是高超技艺，呈现的是客观之美，带给人们的是审美的享受。

⊙The architect/author says: Attention to detail is a sign of the architect's dedication. Through the details we seeks to carry on masterful techniques, by which to present visitors with an aesthetically enjoyable experience.

The Story of JE Mansion

文化意蕴

建筑是文化的载体，没有文化内涵的建筑只是一个空壳。健壹营造的作品最可贵之处，在于用心将建筑营造布置成充满艺术气息的文化居所和空间。

健壹营造从整体到细节都拥有文化主题，并逐渐形成健壹品牌的别致风格。昌平宅院表现的是老北京的府宅韵味，朝阳公馆表现的是堂皇大气的公馆文化，西山景园代表的则是世外桃源般的园林文化。融汇古今、沟通南北、合璧中外的理念除了兼收并蓄的建筑艺术效果之外，也是为了交融汇集古今、南北、中外文化之长。

儒、释、道文化并存于一个建筑作品中：井然有序的空间秩序，是传统儒家文化的典型特征；随宜自然的景物布置，是道家文化的体现；佛堂、佛龛、佛经则是佛教文化的载体。

以健壹景园的佛堂"菩提阁"为例，山门东次间供奉来自东南亚的玉雕佛像以及铜铸四面佛，为典型的南传佛教空间；西次间供奉释迦牟尼、药师和阿弥陀佛三座佛像以及观音菩萨和地藏王菩萨像，为汉传佛教空间。而主院为藏传佛教空间，正殿供奉释迦牟尼佛以及长寿佛、阿弥陀佛、绿度母、文殊菩萨，墙上悬藏传佛教格鲁派（黄教）宗喀巴大师、宁玛派（红教）莲花生大师、萨迦派（花教）五祖、噶举派（白教）玛尔巴译师唐卡画像，东壁供奉萨迦法王照片，西壁书柜陈列藏文版《大藏经》；东配殿为财神殿，供奉财宝天王、黄财神、白财神三位神像；西配殿为护法殿，供奉大黑天、婆罗相吉祥估主、木头明王金刚三尊神像。殿内墙壁均刷为红色，悬藏式经幡，陈列藏式法器。佛教三大流派的文化特质荟萃一地，佛光普照，圆融明净。

中国古典建筑和园林的文化主题不但通过空间格局、立面造型和园林景观来表达，还通过建筑装饰细节的匾额、楹联、书画、陈设来加以提示和渲染。构思新颖、措辞佳妙、含义深刻的匾联，不但可以直接体现主人或设计者良好的文化修养，同时也以文学的方式对园林景致进行了点睛式的渲染和提示，由此成为空间意境塑造的重要手段；匾联本身大多是优秀的书法作品，同样可为景致增色。正如中国古典文学名著《红楼梦》中所写："偌大景致，若干亭榭，无字标题，也觉寥落无趣，任有花柳山水，也断不能生色。"如果把建筑比作立体的国画，那么匾额、楹联正如画上的题款，不可或缺。健壹营造中拥有大量的匾额和楹联，提示每一空间的文化主题，雅俗共赏，耐人寻味。

昌平宅院的餐聚格子间，均以传统乐器为名：舞笙、抚琴、鼓瑟、醒笛、醉箫、亮铍、酣鼓、远钟；戏楼上的独立厅房借著名京戏剧目命名：西厢记、群英会、定军山、三岔口、将相和、对花枪、状元媒、望江亭、凤还巢、贵妃醉等。

健壹景园书苑的包间，也以古典戏曲的长生殿、六义园、桃花扇、春江月、锁麟囊、牡丹亭、玉蜻蜓、琵琶记、诗文会、珍珠记等曲目为名，浓郁的戏剧氛围和古乐格调，萦绕于书苑之间。此外，雅集宴客的厅房名称：三省堂、天府阁、国子园、三友轩、麒麟厅、和合堂、天长阁、景福轩诸名，无不蕴含中国传统文化典故或经典寓意。

健壹建筑中大多数室内的空间都悬挂有字画，几乎都是古代和近现代名家的杰作，部分为原作，部分为临摹精品。其中名画家于非闇先生和京剧大师梅兰芳先生的画作数量最多。书法作品除了匾额、楹联之外，以古代法帖和手抄佛经为主，其中弘一大师所书的《金刚经》《心经》尤为珍贵。

所有作宾健壹建筑的人都会有一个直观又深刻的印象：书盈四壁，坐拥万卷。书柜、书架、书桌几乎无处不在，所藏图书数以万计；客人随时随地可以坐下，一册书卷在手，一杯清茗做伴，闲度半日时光。浓郁的书卷之气，也为健壹营造凭添了无形的文化氛围。

Cultural Theme

Architecture is a cultural expression, and without culture a building is merely a shell. One valuable aspect of JE's works is that they go to great lengths to fashion each structure into an embodiment of culture and artistry.

Each of JE's projects, from overall design to the smallest detail, is based on a single cultural theme, and this has gradually become a part of the team's characteristic style. The Changping Manor is a recreation of an Old Beijing luxury home, the Chaoyang Mansion is infused with the bold elements of an imperial palace, while the West Mountain Garden represents the culture of classical Chinese gardens. Aside from these main themes, many elements of old and new, north and south, and east and west are also fused within.

Elements of the three main religions of China can also be seen: orderly and economic spaces are characteristic of Confucian culture; naturally scattered articles reflect Daoist thought; and Buddhist halls, altars and scriptures also abound.

One particularly notable example is the Bodhi Temple of the JE Garden, which includes characteristics of the three major branches of Buddhism, and features a distinctively lit space.

The cultural theme of classical Chinese architecture is expressed not only through space, structural components and garden scenery, additions such as calligraphy tablets, paintings and furnishings are also essential. These details not only convey the cultural cultivation of the owner or designer, they also serve as an important means to completing the cultural atmosphere of the spaces.

The private dining rooms of the Changping Manor are all named after traditional Chinese musical instruments, while the names of the private opera viewing rooms upstairs are based on Beijing opera pieces.

The rooms of the JE Garden library are also named after opera pieces, adding to the refined cultural atmosphere of the spaces.

书籍静静地在每一个角落等待与您邂逅

Books lie quietly in every corner, patiently waiting to be discovered

The Story of JE Mansion

品酒大厅开启佳酿密码
The spacious banquet hall serves for a lively social get-together

以匾额、楹联来渲染主题
Horizontal and vertical wooden tablets add to the overall theme

许多老匾额属于珍贵文物
Many of the old tablets found are valuable antiques

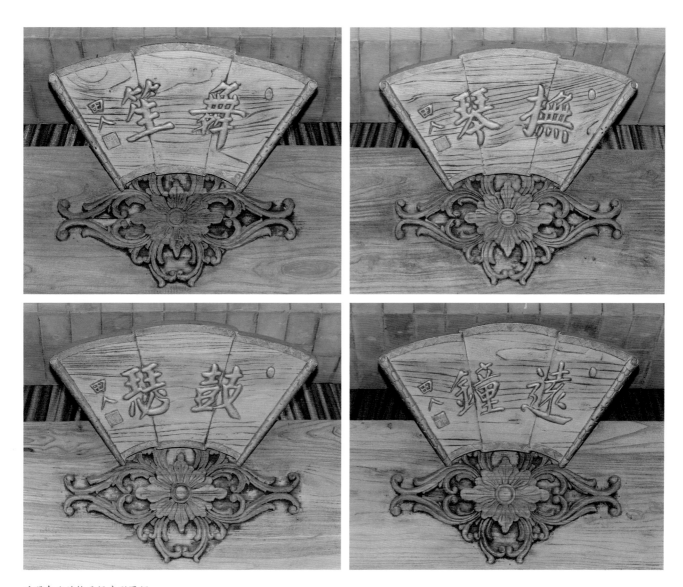

乐器命名的格子间扇形匾额
Private rooms marked with names of instruments on fan-shaped wood carvings

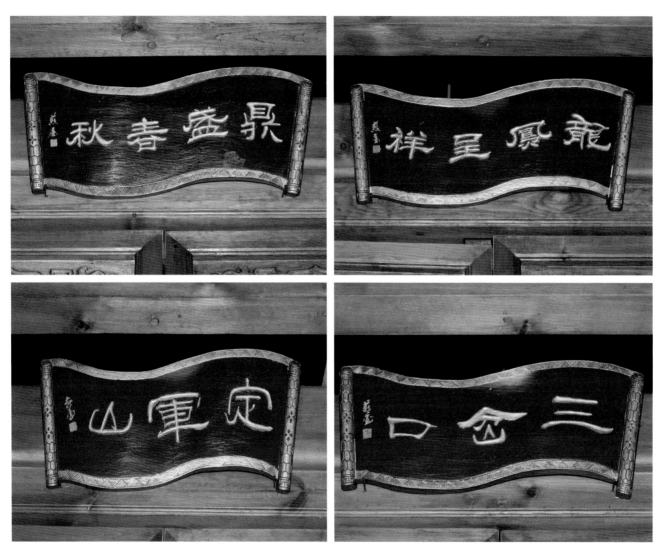

包房匾额以剧目命名
Private rooms are named after opera pieces

佛教文化空间
Spaces featuring Buddhist culture themes

现代雕塑表现传统京剧文化
Modern sculptures convey traditional Beijing opera culture

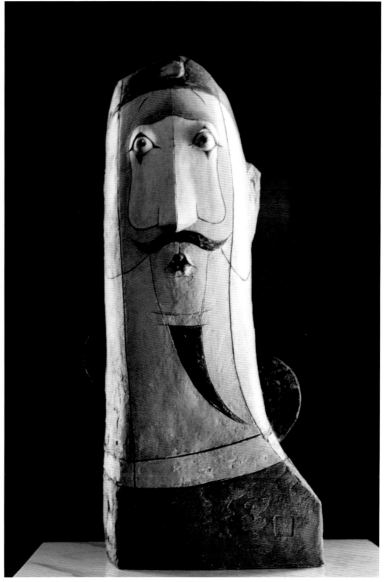

砖雕
Brick carvings

现代雕塑《千里走单骑》
A modern sculpture, "Riding Alone for Thousands of Miles"

现代工艺打造的书法背屏

A calligraphy screen created through modern methods

◎筑(著)者说：我们所做的并不是简单的复古，而是追求独特并充满细节。来访的顾客期待的是享受，而享受的感觉是超越地域时空，且跨越不同文化的。

⊙The architect/author says: What we do is not simple restoration. We strive to achieve uniqueness and immaculate detail. What visitors seek is enjoyment, and the sense of enjoyment spans not only time and space, but culture as well.

筑◎著◎住◎注◎驻

我们设计营造的建筑，是为人们居住和享受的，所以讲究的是人与环境之间的互动与兼容。希望朋友们在我们的建筑里见识到赏心悦目的风景，体会到诗情画意的生活趣味。

十余年，三座建筑，三个作品，倾注了我本人和团队伙伴们的信念、追求、努力和匠心。这些建筑对于我个人而言，仿佛自己人生大书的创作轨迹，循序渐进，发生衍展；每一座建筑的落成与运营，都是对自己、对团队、对环境他人的一个交代。

当我每次驻足在自己设计和建造的院子里、房间中、亭子下、廊桥边、花园里、影壁前，或是和亲朋好友们驻留其中共度人生美好时刻的那些瞬间，时常会有这样的感怀和念头：作为一名设计者和建造者，多年过后，我们这些盖房子修建筑的建设者可能已经不在，但这些建筑或许还存在；或许还有后人在这些建筑里面流连驻足，体会谈论我们的这些留痕与故事，想来也是极欣慰和蛮有趣的……

钱锺书先生有本书叫作《写在人生边上》，说是人生仿佛一本大书，我们人类既是著者又是读者。我想我们这三座建筑作品就是我自己的人生大书，记录的都是自己人生发展的历练轨迹，是对环境生活的注释与眉批。真挚地欢迎各位朋友得空时，到宅院来，到公馆来，到景园来，让我们一同驻足品味，一起赏析阅读。

The architect/author/designer/creator/cultural inheritor says:
Design, Construct, Inspire

The complexes we built are to be lived in and enjoyed, so we focus on interaction and integration between people and environments. We hope that our visitors are captivated by what they see, and that they experience a style of living like something out of a classical painting.

Over the past decade or so, we have created three installments in our series. Each of these embodies the beliefs, aspirations, determination and spirit of myself and the rest of the JE team. To me these structures are chapters in the story of my life, each one breaking new ground and presenting the results of new explorations. The completion and operation of each of these complexes serve as rewards to myself, my team, and our world.

When we walk throughout the courtyards, rooms, corridors and gardens that we have created, or stop to enjoy a moment with close friends amongst the pleasant scenery, I often think that, many years from now, when I'm no longer around, these buildings may still stand there, so I hope that countless others may enjoy them, and discuss the stories they hold within. I find this both comforting and amusing to think about.

Mr. Qian Zhongshu wrote a book called *The Marginalia of Life*, in which he said that life is much like a big book, of which we are both author and reader. I believe that these three complexes are the big book of my life, recording all the experiences and challenges that I've had. I sincerely welcome all to visit the Manor, the Mansion and the Garden, so that we may take a moment to read a chapter or two together.

健 壹
営造志

The Story of
JE Mansion